T0252243

ANNALS OF MATHEMATICS STUDIES

Number 26

ANNALS OF MATHEMATICS STUDIES

Edited by Emil Artin and Marston Morse

7. Finite Dimensional Vector Spaces, *by* Paul R. Halmos
11. Introduction to Nonlinear Mechanics, *by* N. Kryloff *and* N. Bogoliuboff
14. Lectures on Differential Equations, *by* Solomon Lefschetz
15. Topological Methods in the Theory of Functions of a Complex Variable, *by* Marston Morse
16. Transcendental Numbers, *by* Carl Ludwig Siegel
17. Problème Général de la Stabilité du Mouvement, *by* M. A. Liapounoff
18. A Unified Theory of Special Functions, *by* C. A. Truesdell
19. Fourier Transforms, *by* S. Bochner *and* K. Chandrasekharan
20. Contributions to the Theory of Nonlinear Oscillations, *edited by* S. Lefschetz
21. Functional Operators, Vol. I, *by* John von Neumann
22. Functional Operators, Vol. II, *by* John von Neumann
23. Existence Theorems in Partial Differential Equations, *by* Dorothy Bernstein
24. Contributions to the Theory of Games, *edited by* A. W. Tucker
25. Contributions to Fourier Analysis, *by* A. Zygmund, W. Transue, M. Morse, A. P. Calderon, *and* S. Bochner
26. A Theory of Cross-Spaces, *by* Robert Schatten
27. Isoperimetric Inequalities in Mathematical Physics, *by* G. Polya *and* G. Szego

A THEORY OF CROSS-SPACES

BY ROBERT SCHATTEN

PRINCETON

PRINCETON UNIVERSITY PRESS

1950

Printed in the United States of America
by Westview Press, Boulder, Colorado

Princeton University Press On Demand Edition, 1985

TABLE OF CONTENTS

A THEORY OF CROSS-SPACES

INTRODUCTION

1. Statement of the problem.

For two linear vector spaces $\mathfrak{R}_1$ and $\mathfrak{R}_2$, the well-known concept of their "direct sum" in symbol $\mathfrak{R}_1 \oplus \mathfrak{R}_2$, is defined in perfect generality as the linear space of all pairs (f , g) where $f \in \mathfrak{R}_1$, $g \in \mathfrak{R}_2$, with the stipulations:

$$(f_1 , g_1) + (f_2 , g_2) = (f_1 + f_2 , g_1 + g_2) ,$$

$$a(f , g) = (af , ag) \quad \text{for any constant } a .$$

Identifying $(f , 0)$ with f and $(0 , g)$ with g, we may assume that $\mathfrak{R}_1$ and $\mathfrak{R}_2$ are subspaces of $\mathfrak{R}_1 \oplus \mathfrak{R}_2$. Thus, $f + g$ is defined and the elementary rules of addition hold.

When $\mathfrak{B}_1$ and $\mathfrak{B}_2$ represent two Banach spaces, it is customary to introduce a norm α on $\mathfrak{B}_1 \oplus \mathfrak{B}_2$ in a manner which assures that the sequence of pairs (f_n , g_n) is convergent in that norm to (f , g) if, and only if, the sequence $f_n \rightarrow f$ and $g_n \rightarrow g$ in the Banach spaces $\mathfrak{B}_1$ and $\mathfrak{B}_2$ respectively. With the last condition, the resulting "direct sum" $\mathfrak{B}_1 \oplus_\alpha \mathfrak{B}_2$ is complete, hence a Banach space. Together with $\mathfrak{B}_1$, $\mathfrak{B}_2$, we consider their conjugate spaces $\mathfrak{B}_1^*$, $\mathfrak{B}_2^*$, and form the linear space $\mathfrak{B}_1^* \oplus \mathfrak{B}_2^*$ of pairs (F , G) for $F \in \mathfrak{B}_1^*$, $G \in \mathfrak{B}_2^*$. Clearly, an additive bounded functional $\mathfrak{F}$ on $\mathfrak{B}_1 \oplus_\alpha \mathfrak{B}_2$ corresponds to a unique pair (F , G) and conversely, subject to the relationship $\mathfrak{F}(f , g) = \mathfrak{F}(f , 0) + \mathfrak{F}(0 , g) = F(f) + G(g)$. This correspondence is clearly additive. The bound of such a functional $\alpha'(\mathfrak{F}) = \alpha'(F, G)$ determines a norm α' on $\mathfrak{B}_1^* \oplus \mathfrak{B}_2^*$ and

$$(\mathfrak{B}_1 \oplus_\alpha \mathfrak{B}_2)^* = \mathfrak{B}_1^* \oplus_{\alpha'} \mathfrak{B}_2^*$$

always holds.

When $\mathfrak{H}_1$ and $\mathfrak{H}_2$ are Hilbert spaces, $\mathfrak{H}_1 \oplus \mathfrak{H}_2$ becomes also a Hilbert space if the inner product of the pairs (f_1, g_1), (f_2, g_2) is defined by the number $(f_1, f_2) + (g_1, g_2)$.

For Hilbert spaces the operation $\oplus$ has turned out to be a powerful tool quite often used, for instance, in the theory of closed and adjoint operators.

The theory of the direct sum of two Banach spaces and in particular of two Hilbert spaces does not present any special difficulties and is generally well-known.

These considerations suggest the following natural question: Given two linear vector spaces $\mathfrak{R}_1$ and $\mathfrak{R}_2$, is it possible to construct a "direct product" that is a linear vector space say denoted with the symbol $\mathfrak{R}_1 \otimes \mathfrak{R}_2$ of formal products $f \otimes g$, for which multiplication laws, that is, the distributive laws

$$(f_1 + f_2) \otimes g = f_1 \otimes g + f_2 \otimes g,$$

$$f \otimes (g_1 + g_2) = f \otimes g_1 + f \otimes g_2,$$

hold?.

Furthermore, when $\mathfrak{B}_1$ and $\mathfrak{B}_2$ are Banach spaces, we would demand that a suitable (one or more) norm α on $\mathfrak{B}_1 \otimes \mathfrak{B}_2$ may be so defined, as to make $\mathfrak{B}_1 \otimes_\alpha \mathfrak{B}_2$ a Banach space. Moreover, the elements of $\mathfrak{B}_1^* \otimes \mathfrak{B}_2^*$ should determine the additive bounded functionals on $\mathfrak{B}_1 \otimes_\alpha \mathfrak{B}_2$. Their bounds should furnish a conjugate norm α' on $\mathfrak{B}_1^* \otimes \mathfrak{B}_2^*$ and the relationship

$$(\mathfrak{B}_1 \otimes_\alpha \mathfrak{B}_2)^* = \mathfrak{B}_1^* \otimes_{\alpha'} \mathfrak{B}_2^*$$

should hold.

In particular, when $\mathfrak{H}_1$ and $\mathfrak{H}_2$ are Hilbert spaces, we are able to define an inner product on $\mathfrak{H}_1 \otimes \mathfrak{H}_2$ so as to make it also a Hilbert space.

Finally we should be interested in finding out whether the operation $\otimes$ so defined presents similar possibilities as $\oplus$?

It is the purpose of this exposition to present a reasonable solution of this problem.

2. Purpose of this exposition.

From the algebraic standpoint alone the notion of the "direct product" for two finite dimensional linear vector spaces has been in a sense mentioned by H. Weyl [25]. Let $\mathfrak{E}_1$ denote the space of m-tuples of numbers $(x_1, \ldots, x_m)$ and let $\mathfrak{E}_2$ denote the space of n-tuples $(y_1, \ldots, y_n)$. The "product" $(x_1, \ldots, x_m) \otimes (y_1, \ldots, y_n)$ is defined as the mn-tuple $(x_1 y_1, \ldots, x_m y_1; \ldots; x_1 y_n, \ldots, x_m y_n)$. For these products the distributive laws hold. The products alone do not form a linear manifold. However, the set of their finite sums, which we shall denote by $\mathfrak{E}_1 \otimes \mathfrak{E}_2$, fills the entire mn dimensional space. Such a product may be interpreted as a matrix of rank ≤ 1 with m rows and n columns, hence also as an operator. Thus for finite dimensional spaces, $\mathfrak{E}_1 \otimes \mathfrak{E}_2$ represents essentially the space of all rectangular matrices with a fixed number of rows and columns, hence it possesses a well-known algebraic structure.

But even when the linear spaces $\mathfrak{R}_1$, $\mathfrak{R}_2$, are infinite dimensional, a linear space $\mathfrak{R}_1 \otimes \mathfrak{R}_2$ may be constructed in an elementary manner (as the totality of finite sums of "products"). The algebraic aspects of the $\otimes$ operation for general linear spaces (even for abelian groups) have been discussed successfully by H. Whitney [26]. $\mathfrak{R}_1 \otimes \mathfrak{R}_2$ for two algebras has been defined by J. L. Dorroh [3]. Later it has been also occasionally employed by F. J. Murray in [7] and [8], in dealing with finite dimensional l_p spaces, and in Hilbert spaces.

While the definition of the linear set $\mathfrak{R}_1 \otimes \mathfrak{R}_2$ is quite simple, the serious problem centers around the possible topologies and in particular, the possible norms that can be introduced in $\mathfrak{B}_1 \otimes \mathfrak{B}_2$, when $\mathfrak{B}_1$ and $\mathfrak{B}_2$ represent Banach spaces. This is true even for the case when both spaces are finite dimensional. In this connection it should be mentioned that an interesting class of norms for the space of square matrices of n-th order, has been defined by J. von Neumann [23]. We shall have occasion later to discuss his result. For two Hilbert spaces $\mathfrak{H}_1$ and $\mathfrak{H}_2$ a natural definition of an "inner product" in $\mathfrak{H}_1 \otimes \mathfrak{H}_2$ has been readily derived in [9]. Furthermore, the discussion for an infinite direct product of Hilbert spaces has been carried out by J. von Neumann [19].

Thus, so far the study of $\mathfrak{B}_1 \otimes \mathfrak{B}_2$ even in its elementary aspects were incomplete and restricted to a few special spaces only, in which it was always assumed the existence of a at most denumerable basis or an inner product. In each of these cases the considerations were restricted to a <u>single</u> special norm.

It is the purpose of this exposition to present the first attempt of the study of the theory of cross-spaces in its general form, for perfectly general

Banach spaces. It turns out that the infinite dimensional case displays some interesting and unexpected phenomena which present some new difficulties. These new phenomena, of which let us just mention the distinction between an associate and conjugate space for a given cross-space, are mostly responsible for the growth of the present theory in a natural manner, which also includes the partially known theory of crossing two finite-dimensional spaces.

Although some of the results concerning this problem have been published before in the following sequence of papers [23] , [12] , [13] , [14] , [4] , [15] , [17] , [16] quoted in the chronological order of their appearance, yet, the present treatment is desirable for the following reason: The seemingly innocent --at the start--problem of "crossing" two Banach spaces, gradually grew into an extensive field with interesting applications. Accordingly, definitions, statements and notation (which presents quite a problem in itself) had to be revised from time to time to suit the new needs. The present exposition also includes most of these scattered results in a unified theory. A few results herein, have not been published before. Wherever possible, the published ones have been refined.

While the present theory--which has turned out to furnish an effective tool in dealing with Banach spaces whose elements are operators on some Banach space--is in an advanced stage, it is far from being complete. As a matter of fact a number of interesting problems are still open. Some are mentioned in the body of this paper, and the main difficulty in their solution is pointed out. It is hoped that this presentation will induce the interested reader to further investigations in this promising field.

3. Acknowledgement.

At this point it seems proper to acknowledge that the author's discussions with Professor J. von Neumann in 1944-46 (during the author's membership at The Institute for Advanced Study) followed up by an exchange of correspondence in 1946-48 have played a decisive part in preparing the foundation for this draft. Needless to say that the ideas contained in this draft were originated by both Professor von Neumann and the author. Some of these, were published (although in different form) in their joint papers [15] and [16]. While the authors assume full responsibility for the shortcomings that may be contained in this exposition, the merits and credits it may have must be shared with Professor von Neumann.

4. Plan of study.

To have some idea what to expect, suppose first that $\mathfrak{B}_1$ and $\mathfrak{B}_2$ are finite, say p and q dimensional Banach spaces, whose elements will be denoted by f and g, while $\mathfrak{B}_1^*$ and $\mathfrak{B}_2^*$ denote their conjugate spaces whose elements we denote by F and G.

As was mentioned before, we may then interpret $f \otimes g$ for instance, as the operator $F(f)g$ from $\mathfrak{B}_1^*$ into $\mathfrak{B}_2$ (or $G(g)f$ from $\mathfrak{B}_2^*$ into $\mathfrak{B}_1$) of rank $\leqslant 1$. The expression $\sum_{i=1}^n f_i \otimes g_i$ may be interpreted similarly, as the operator $\sum_{i=1}^n F(f_i)g_i$ from $\mathfrak{B}_1^*$ into $\mathfrak{B}_2$. Furthermore, every operator from $\mathfrak{B}_1^*$ into $\mathfrak{B}_2$ can be obtained in such a manner. The algebra for these expressions and in particular the distributive laws of the symbol $\otimes$ are determined by the obvious algebraic laws for operators. We denote by $\mathfrak{B}_1 \otimes \mathfrak{B}_2$ the linear space of such expressions. The last may clearly be identified with the pq dimensional linear space of operators from $\mathfrak{B}_1^*$ into $\mathfrak{B}_2$.

Defining on it a norm α (this can be done in many ways) we obtain a "direct product" $\mathfrak{B}_1 \otimes_\alpha \mathfrak{B}_2$. The last is also termed a "cross-space" whenever α is a "cross-norm" that is, $\alpha(f \otimes g) = \|f\| \|g\|$. Similarly, we construct the linear space $\mathfrak{B}_1^* \otimes \mathfrak{B}_2^*$ of expressions $\Sigma_{j=1}^m F_j \otimes G_j$. For a fixed $\Sigma_{j=1}^m F_j \otimes G_j$ the double sum $\Sigma_{j=1}^m \Sigma_{i=1}^n F_j(f_i)G_j(g_i)$ represents an additive bounded functional of expressions on $\mathfrak{B}_1 \otimes_\alpha \mathfrak{B}_2$. Its bound $\alpha'(\Sigma_{j=1}^m F_j \otimes G_j)$ determines a norm α' on $\mathfrak{B}_1^* \otimes \mathfrak{B}_2^*$ "associated" with α. Moreover, since every additive bounded functional can be obtained in such a manner, we have $(\mathfrak{B}_1 \otimes_\alpha \mathfrak{B}_2)^* = \mathfrak{B}_1^* \otimes_{\alpha'} \mathfrak{B}_2^*$.

Guided by these considerations we define a direct product for general Banach spaces. The situation is a trifle simpler for reflexive spaces. In that case again as before, the linear space of "expressions" may be interpreted as the precise space of operators of <u>finite</u> <u>rank</u> from $\mathfrak{B}_1^*$ into $\mathfrak{B}_2$. For the non-reflexive case however, the set of expressions determines a proper subspace of the linear space of operators of finite rank. In that case the most general approach is desirable by considering "formal products" and "formal expressions" subject to the rules suggested by the previous case. Once a linear set of expressions is constructed, then following the previous pattern we define on it a norm α. In general however, the resulting normed linear space may not be complete. Whenever this occurs we "complete" it, that is, imbed it in the usual Cantor-Meray fashion in the smallest possible Banach space which we shall denote by $\mathfrak{B}_1 \otimes_\alpha \mathfrak{B}_2$. As before the norm α determines an associate norm α' on the linear space of expressions $\Sigma_{j=1}^m F_j \otimes G_j$. The last space when completed furnishes the "associate space" $\mathfrak{B}_1^* \otimes_{\alpha'} \mathfrak{B}_2^*$. Here however, we are already confronted with the following interesting phenomenon. The conjugate space for a given cross-space may (and generally does) contain the associate space as a proper subspace. Finding out their exact relationship as well as the characterization of some of the cross-spaces their associate and conjugate spaces, is one of the main problems of the present exposition.

5. Outline of results.

In Chapter I, we consider two Banach spaces $\mathfrak{B}_1$, $\mathfrak{B}_2$, without any special restrictions. For $f \in \mathfrak{B}_1$, $g \in \mathfrak{B}_2$, we construct "formal products" $f \otimes g$. With these, we form a linear set $\mathfrak{B}_1 \odot \mathfrak{B}_2$ of "formal expressions" (that is, finite sums) $\sum_{i=1}^{n} f_i \otimes g_i$. Although these expressions are abstract algebraic elements, yet we shall often use for them a specific representation by means of "operators of finite rank". Indeed, an expression $\sum_{i=1}^{n} f_i \otimes g_i$ may be interpreted an an operator A from $\mathfrak{B}_1^*$ into $\mathfrak{B}_2$ of finite rank, defined by means of the relation $A(F) = \sum_{i=1}^{n} F(f_i)g_i$. In particular, expressions which furnish the same operator are considered "equivalent" and are combined into a single element. For these expressions we set up algebraic rules as fo example, the one expressing the distributive laws of the $\otimes$ symbol, addition of expressions and multiplication of scalars by expressions. These are suggested by the algebraic laws governing operators. Next we consider some algebraic relationships between these expressions, since the distributive property of the $\otimes$ symbol, introduces certain linear dependencies.

Together with $\mathfrak{B}_1 \odot \mathfrak{B}_2$ we similarly form the linear set $\mathfrak{B}_1^* \odot \mathfrak{B}_2^*$ of expressions $\sum_{j=1}^{m} F_j \otimes G_j$ (for $F_j \in \mathfrak{B}_1^*$ and $G_j \in \mathfrak{B}_2^*$). For a pair of expressions $\sum_{i=1}^{n} f_i \otimes g_i$ and $\sum_{j=1}^{m} F_j \otimes G_j$ an "inner product" $(\sum_{j=1}^{m} F_j \otimes G_j)(\sum_{i=1}^{n} f_i \otimes g_i)$ is defined by means of the number $\sum_{j=1}^{m} \sum_{i=1}^{n} F_j(f_i) G_j(g_i)$ whose invariance under "equivalence" is proven.

In Chapter II, we define in the usual fashion a "norm" $\alpha(\sum_{i=1}^{n} f_i \otimes g_i)$ on $\mathfrak{B}_1 \odot \mathfrak{B}_2$ and obtain a normed linear space $\mathfrak{B}_1 \odot_\alpha \mathfrak{B}_2$. This can naturally be done in many ways. Of interest however are the "crossnorms", that is, those which satisfy the condition $\alpha(f \otimes g) = \| f \| \, \| g \|$ for any pair f, g. Among the last ones we single out those which satisfy the "uniformity condition", that is, $\alpha(\sum_{i=1}^{n} Sf_i \otimes Tg_i) \leqslant |||S||| \, |||T||| \, \alpha(\sum_{i=1}^{n} f_i \otimes g_i)$ for any pair of operators S and T on $\mathfrak{B}_1$ and $\mathfrak{B}_2$ respectively. The significance of this class is discussed later.

For a given norm α on $\mathfrak{B}_1 \odot \mathfrak{B}_2$ we construct an "associate" norm on $\mathfrak{B}_1^* \odot \mathfrak{B}_2^*$, defining $\alpha'(\sum_{j=1}^{m} F_j \otimes G_j)$ as the bound of the additive functional $(\sum_{j=1}^{m} F_j \otimes G_j)(\sum_{i=1}^{n} f_i \otimes g_i)$ of expressions $\sum_{i=1}^{n} f_i \otimes g_i$ on $\mathfrak{B}_1 \odot_\alpha \mathfrak{B}_2$.

Due to their significance two uniform crossnorms are singled out and discussed in detail, namely, the greatest crossnorm γ and the bound λ. We show that for any two Banach spaces, a unique greatest crossnorm can be always constructed. The last one is therefore of "general character". So is also the crossnorm λ, where $\lambda(\sum_{i=1}^{n} f_i \otimes g_i)$ represents the bound of the operator $\sum_{i=1}^{n} F(f_i) g_i$ from $\mathfrak{B}_1^*$ into $\mathfrak{B}_2$. Moreover, λ is also of "local character", that is, the value of $\lambda(\sum_{i=1}^{n} f_i \otimes g_i)$ depends only on the spatial relations between the f_i's and those between the g_i's, and not on the including them spaces. We establish that λ is the least crossnorm whose associate λ' is also a crossnorm. Furthermore, λ when defined on $\mathfrak{B}_1^* \odot \mathfrak{B}_2^*$ may be considered as the associate with the greatest crossnorm γ defined on $\mathfrak{B}_1 \odot \mathfrak{B}_2$. Finally, we prove that whenever for a crossnorm α its "first associate" α' is also a crossnorm, then its associates of higher order, that is, α'', α''',..... are also crossnorms.

In Chapter III, we "complete" the normed linear space $\mathfrak{B}_1 \odot_\alpha \mathfrak{B}_2$ in the usual Cantor-Meray fashion [6, p. 106], that is, imbed it into the smallest possible Banach space by considering the space of fundamental (Cauchy) sequences of expressions (that is, elements which they represent) in $\mathfrak{B}_1 \odot_\alpha \mathfrak{B}_2$, and introducing some standard identifications. The completed space is a Banach space which we shall denote by $\mathfrak{B}_1 \otimes_\alpha \mathfrak{B}_2$. The last space which naturally depends on α is defined as a "direct product" of $\mathfrak{B}_1$ and $\mathfrak{B}_2$, or whenever α is a crossnorm also as a "cross-space".

For a crossnorm $\alpha \geqq \lambda$ on $\mathfrak{B}_1 \odot \mathfrak{B}_2$, the α' is also a crossnorm on $\mathfrak{B}_1^* \odot \mathfrak{B}_2^*$. The finite number $\alpha'(\sum_{j=1}^{m} F_j \otimes G_j)$ represents the bound of the additive functional $(\sum_{j=1}^{m} F_j \otimes G_j)(\sum_{i=1}^{n} f_i \otimes g_i)$ on $\mathfrak{B}_1 \odot_\alpha \mathfrak{B}_2$, therefore also on $\mathfrak{B}_1 \otimes_\alpha \mathfrak{B}_2$. Completing $\mathfrak{B}_1^* \odot_{\alpha'} \mathfrak{B}_2^*$ we obtain the "associate space" $\mathfrak{B}_1^* \otimes_{\alpha'} \mathfrak{B}_2^*$. Thus, $(\mathfrak{B}_1 \otimes_\alpha \mathfrak{B}_2)^* \supset \mathfrak{B}_1^* \otimes_{\alpha'} \mathfrak{B}_2^*$. The last inclusion depending on the particular Banach spaces $\mathfrak{B}_1$, $\mathfrak{B}_2$, and crossnorm α under consideration, is in many cases a proper one. Therefore, for a given crossnorm α, the cross-space $\mathfrak{B}_1 \otimes_\alpha \mathfrak{B}_2$ determines uniquely a conjugate space $(\mathfrak{B}_1 \otimes_\alpha \mathfrak{B}_2)^*$ and an associate space $\mathfrak{B}_1^* \otimes_{\alpha'} \mathfrak{B}_2^*$. Although we have a fairly good idea what the associate space of a given cross-space represents, we do not find it easy to state the precise restrictions imposed on a crossnorm α for which the resulting cross-space is such, that its conjugate space coincides with its associate space. A complete discussion has been presented however, for the case when we deal with the greatest crossnorm γ . In the last case, $(\mathfrak{B}_1 \otimes_\gamma \mathfrak{B}_2)^*$ may be characterized as the Banach space of all operators from $\mathfrak{B}_1$ into $\mathfrak{B}_2^*$ (from $\mathfrak{B}_2$ into $\mathfrak{B}_1^*$) where the bound of an operator represents its norm. On the other hand, $\mathfrak{B}_1^* \otimes_{\gamma'} \mathfrak{B}_2^* = \mathfrak{B}_1^* \otimes_\lambda \mathfrak{B}_2^*$ may be interpreted as the Banach space of all operators from $\mathfrak{B}_1$ into $\mathfrak{B}_2^*$ (from $\mathfrak{B}_2$ into $\mathfrak{B}_1^*$), which may be approximated "in bound" by operators of finite rank. Incidentally, we establish a "natural equivalence" between the space of all (those which can be approximated "in bound" by finite rank) operators from $\mathfrak{B}_1$ into $\mathfrak{B}_2^*$ and the space of all (those which can be approximated "in bound" by finite rank) operators from $\mathfrak{B}_2$ into $\mathfrak{B}_1^*$.

Let α be a given crossnorm on $\mathfrak{B}_1 \odot \mathfrak{B}_2$. An operator A from $\mathfrak{B}_1$ into $\mathfrak{B}_2^*$ is said to be of "finite α-norm" whenever there exists a finite constant C such that $|\Sigma_{i=1}^n (Af_i)g_i| \leqslant C\alpha(\Sigma_{i=1}^n f_i \otimes g_i)$ for every expression $\Sigma_{i=1}^n f_i \otimes g_i$. The least of such constants is denoted by $\|A\|_\alpha$. We always have $\|A\|_\alpha \geqslant |||A|||$. For any crossnorm $\alpha \geqslant \lambda$, the space of operators A from $\mathfrak{B}_1$ into $\mathfrak{B}_2^*$ of finite α-norm is a normed linear space if $\|A\|_\alpha$ represents the norm of A . This space is complete. Moreover, it may be interpreted as $(\mathfrak{B}_1 \otimes_\alpha \mathfrak{B}_2)^*$. On the other hand, $\mathfrak{B}_1^* \otimes_{\alpha'} \mathfrak{B}_2^*$ may be considered as the Banach space of all operators from $\mathfrak{B}_1$ into $\mathfrak{B}_2^*$ (from $\mathfrak{B}_2$ into $\mathfrak{B}_1^*$) of finite α-norm which may be approximated in that norm by operators of finite rank. For every operator A we have $\|A\|_\gamma = |||A|||$.

Finally we settle the "extension" problem for γ by proving that in general γ is not of "local character". A Banach space $\mathfrak{B}$ is unitary if and only if, $\mathfrak{M} \otimes_\gamma \mathfrak{M}^* \subset \mathfrak{B} \otimes_\gamma \mathfrak{M}^*$ for every two-dimensional linear manifold $\mathfrak{M} \subset \mathfrak{B}$.

In Chapter IV, a Banach space $\boldsymbol{\mathfrak{B}}$ of operators from $\mathfrak{B}_1$ into $\mathfrak{B}_2^*$ is termed an "ideal" if, (i) together with $A \in \boldsymbol{\mathfrak{B}}$ also $YAX \in \boldsymbol{\mathfrak{B}}$ for any pair of operators X and Y on $\mathfrak{B}_1$ and $\mathfrak{B}_2^*$ respectively, and (ii) $\| YAX \| \leqslant |\!|\!| X |\!|\!| \; |\!|\!| Y |\!|\!| \; \| A \|$, where $\| A \|$ stands for the norm of A in $\boldsymbol{\mathfrak{B}}$. For a crossnorm α, the Banach space of operators of finite α-norm, that is, $(\mathfrak{B}_1 \otimes_\alpha \mathfrak{B}_2)^*$ is an ideal if and only if, α is uniform. If α is uniform, α' is such, and $\mathfrak{B}_1^* \otimes_{\alpha'} \mathfrak{B}_2^*$ forms an ideal.

In Chapter V, we make use of our previous results and avail ourselves in addition of the tools at our disposal in unitary spaces $\mathfrak{H}$ (in the sense, finite dimensional Euclidean spaces or Hilbert spaces). After some brief remarks on possible interpretations of the direct products of unitary spaces, we decide to interpret $\mathfrak{H} \odot \bar{\mathfrak{H}}$ as the linear space of operators X on $\mathfrak{H}$ of finite rank, and denote a crossnorm accordingly by $\alpha(X)$. We single out the completely continuous operators on $\mathfrak{H}$ and characterize them (hence also the operators of finite rank) as precisely those operators A which may be written uniquely in the "canonical" form $A = \sum_i a_i \varphi_i \otimes \bar{\psi}_i$, where both (φ_i) and (ψ_i) form orthonormal sets, the a_i's are > 0 and $\lim a_i = 0$ whenever the sum has an infinite number of terms. The a_i's form the positive point proper values (with multiplicities) of abs(A).

Following a brief discussion of some properties of symmetric gauge functions, we consider the unitarily invariant crossnorms α on $\mathfrak{H} \odot \bar{\mathfrak{H}}$, that is, those which satisfy the condition $\alpha(UXV^*) = \alpha(X)$ for any operator X of finite rank and any pair of unitary operators U and V. We actually characterize the precise class of unitarily invariant crossnorms as the one which can be generated from the class of symmetric gauge functions on the linear set of n-tuples of real numbers or on the linear set of infinite sequences of real numbers having only a finite number of non-zero terms, depending whether $\mathfrak{H}$ is n-dimensional or a Hilbert space. Furthermore, we prove that the class of unitarily invariant crossnorms coincides with the class of uniform crossnorms, that is, those which satisfy the condition $\alpha(AXB) \leq |||A|||\ |||B|||\ \alpha(X)$ for any operator X of finite rank and any pair of operators A and B. λ represents the least unitarily invariant crossnorm. Consequently, the associate with every unitarily invariant crossnorm is also a unitarily invariant crossnorm. Every unitarily invariant crossnorm α is reflexive, that is, satisfies the condition $\alpha'' = \alpha$ identically.

We define the Schmidt-class (sc) as the class of all those operators A on $\mathfrak{H}$, for which $\sum_i \|A\varphi_i\|^2 < +\infty$, for a complete orthonormal set (φ_i); the last sum is independent on the chosen cnos. The (sc) is a linear set. For A and B in (sc), the sum $\sum_i (A\varphi_i, B\varphi_i) = (A, B)$ is absolutely convergent and independent on the chosen cnos (φ_i). The number (A, B) is an "inner product" on (sc) and $(A, A)^{\frac{1}{2}}$ which we define as $\sigma(A)$ is the norm that goes with it. In particular, $\sigma(X)$ is a crossnorm on the linear set of operators of finite rank. We prove that $\sigma = \sigma'$ and $(\mathfrak{H} \otimes_\sigma \overline{\mathfrak{H}})^* = \mathfrak{H} \otimes_{\sigma'} \overline{\mathfrak{H}}$, that is, the Banach space of all operators of finite σ-norm, may be characterized as the Schmidt-class; they all may be approximated in that norm by operators of finite rank.

Next we show that the linear space of all completely continuous operators on $\mathfrak{H}$ where the bound of an operator is considered as its norm, furnishes the Banach space $\mathfrak{H} \otimes_\lambda \overline{\mathfrak{H}}$. Its first conjugate space coincides with its associate space and may be interpreted as the trace-class. The trace-class is the Banach space of all operators A on $\mathfrak{H}$, for which $\sum_i (\mathrm{abs}(A)\varphi_i, \varphi_i) < \infty$ for a complete orthonormal set (φ_i) and where the last sum which is independent on the chosen complete orthonormal set represents the norm of A. The second conjugate space $(\mathfrak{H} \otimes_\lambda \overline{\mathfrak{H}})^{**}$ may be interpreted as the Banach space of all operators on $\mathfrak{H}$ where again the bound of an operator represents its norm.

In this connection we discuss the topological equivalence of the spaces $\mathfrak{H} \otimes_\lambda \overline{\mathfrak{H}}$ and $\mathfrak{H} \otimes_\alpha \overline{\mathfrak{H}}$ for any "limited" crossnorm α.

A crossnorm α is termed "significant" if every operator of finite α-norm is completely continuous. For any significant unitarily invariant crossnorm $(\mathfrak{H} \otimes_\alpha \overline{\mathfrak{H}})^*$ may be characterized as the Banach space of exactly all those completely continuous operators in the canonical form $\sum_i a_i \varphi_i \otimes \overline{\psi}_i$, for which $\lim_{n \to \infty} \alpha' (\sum_{i=1}^{n} a_i \varphi_i \otimes \overline{\psi}_i) < +\infty$. The value of the last limit represents the norm and equals $\|A\|_\alpha$.

Finally we show that for an n-dimensional space $\mathfrak{H}_n$ with the Euclidean norm (or Hilbert space), λ does not represent the least crossnorm on $\mathfrak{H}_n \odot \mathfrak{H}_n$. The constructed crossnorms which are not $\geq \lambda$ at the same time furnish examples of crossnorms whose associates are not crossnorms. They may also serve as examples of crossnorms which are not unitarily invariant.

In Appendix I, we present some scattered results for further investigations. For the sake of simplicity we assume throughout that $\mathfrak{B}_1$ and $\mathfrak{B}_2$ are reflexive, and all crossnorms α are $\geq \lambda$. We have always, $\alpha'' \leq \alpha$ and $\alpha''' = \alpha'$. The following conditions are equivalent: (i) $\alpha' = \beta'$ implies $\alpha \leq \beta$, (ii) $\alpha'' = \alpha$, (iii) $\alpha = \beta'$ for some β. We term α reflexive whenever (ii) holds. λ is always reflexive.

Next we investigate reflexive cross-spaces generated by means of reflexive crossnorms and prove the equivalence of the following statements: (i) a cross-space is reflexive, (ii) its associate space is reflexive, (iii) the conjugate space of the cross-space coincides with its associate space, and the conjugate space of the associate space coincides with the cross-space. As an example, we also discuss the non-reflexive cross-space $l_p \otimes_\lambda l_q$ where $p > 1$, $\frac{1}{p} + \frac{1}{q} = 1$. In particular both, the trace-class and also the space of all completely continuous operators on a Hilbert space are non-reflexive.

Finally, we introduce "limited" crossnorms and discuss some elementary relations which they satisfy. In particular, they are all reflexive. As a corollary we deduce for instance, that a crossnorm on $\mathfrak{H} \otimes \overline{\mathfrak{H}}$ is not determined by the values it assumes for operators of rank $\leq$ p (where p is any natural number smaller than the dimension of $\mathfrak{H}$).

In Appendix II, we present a definite construction (not unique however), which for any two Banach spaces $\mathfrak{B}_1$, $\mathfrak{B}_2$, (without any special restrictions!), furnishes a definite crossnorm on $\mathfrak{B}_1 \odot \mathfrak{B}_2$. The resulting crossnorm we are justified to term "self-associate" since, when our construction is applied to unitary spaces we obtain the usual self-associate crossnorm σ on $\mathfrak{B}_1 \odot \mathfrak{B}_2$.

We shall assume that the reader is familiar with the elementary concepts and theorems in Banach spaces and in Hilbert spaces, as can be found in [1] and [18].

The definitions and theorems throughout this paper fall into two categories. First, we have theorems which apply to perfectly general (and sometimes only reflexive) Banach spaces, hence equally well to unitary spaces. These we prefer to formulate in the most general form. They form the content of the first four chapters and of both appendices. The other type is formulated only for unitary spaces.

The symbols $\mathfrak{R}_1$, $\mathfrak{R}_2$, will be assigned to two linear spaces in the sense of [1, p. 26], while $\mathfrak{R}_1^*$ and $\mathfrak{R}_2^*$ will stand for the linear space of all additive (which in our terminology will also imply homogeneous) numerically valued functionals [1, p. 27] on $\mathfrak{R}_1$ and $\mathfrak{R}_2$ respectively. $\mathfrak{B}_1$ and $\mathfrak{B}_2$ will stand for two Banach spaces, that is, two normed complete linear spaces [1, p. 53], while $\mathfrak{B}_1^*$ and $\mathfrak{B}_2^*$ will stand for their conjugate spaces [1, p. 188], that is, the Banach spaces of all additive and bounded [1, pp. 54-55] functionals on $\mathfrak{B}_1$ and $\mathfrak{B}_2$ respectively, where the bound of a functional represents its norm. Two Banach spaces will be termed "equivalent" if they can be transformed into each other in a one-to-one additive and norm-preserving fashion, in the sense of [1, p. 180].

$\mathfrak{H}$ will stand for a linear space in which there is defined an inner product (φ, ψ) and which is separable and complete [18, §2]. Thus, $\mathfrak{H}$ is either a Hilbert space or a finite dimensional Euclidean space according to whether it contains an infinite or finite number of linearly independent elements. $\bar{\mathfrak{H}}$ is the space obtained from $\mathfrak{H}$ in which the fundamental operations $a\varphi$, $\varphi+\psi$, (φ, ψ), are replaced by $\bar{a}\varphi$, $\varphi+\psi$, $\overline{(\varphi, \psi)}$. $\mathfrak{M}_1$ and $\mathfrak{M}_2$ will stand for closed linear manifolds in $\mathfrak{B}_1$ and $\mathfrak{B}_2$ or $\mathfrak{H}$.

Elements in $\mathfrak{R}_1$, $\mathfrak{B}_1$ or $\mathfrak{H}$ will be denoted by f, h, f_i, h_i. Those in $\mathfrak{R}_2$, $\mathfrak{B}_2$ or $\mathfrak{H}$ are given by g, k, g_i, k_i. The letters F, H, F_i, H_i, will stand for elements in $\mathfrak{R}_1^*$, $\mathfrak{B}_1^*$, while G, K, G_i, K_i, will be assigned to elements in $\mathfrak{R}_2^*$ and $\mathfrak{B}_2^*$. Elements in $\mathfrak{H}$ will also be represented by φ, ψ and φ_i, ψ_i. By (φ_i), (ψ_j), we shall denote normalized orthogonal sets (nos) or complete normalized orthogonal sets (cnos) in $\mathfrak{H}$, and sometimes in (closed linear) subsets of $\mathfrak{H}$.

The term "operator" will always mean "additive and bounded operator" that is, additive and bounded transformation, whose domain of definition is the whole space and the range is contained in the same or another Banach space. We recall that A is additive if, $A(af + bg) = aAf + bAg$ for any pair of elements f, g, and any constants a, b. The letters A, B, C, X, Y, will be reserved for operators. A^* will denote the adjoint of A in the sense of [1, p. 100] when considered on general Banach spaces. It should be understood however in the sense of [18, Definition 2.8] when considered on unitary spaces. An operator whose range is finite dimensional is termed of "finite rank". The symbol $|||A|||$ will represent the bound of A, while $||A||$ stands for the norm of A, when A is considered as an element of a Banach space whose elements are operators with a norm not necessarily equal to their bound.

For a Hermitean operator A on $\mathfrak{H}$ (that is such that $A^* = A$) we shall use the term "definite" in the sense of non-negative definite, that is, when $(A\varphi, \varphi)$ is always ≥ 0. Unitary operators will be denoted by U, V. We shall reserve the letter W for a partially isometric operator on $\mathfrak{H}$, that is, one which is isometric on a closed linear manifold $\mathfrak{M}$ and equal to 0 on its orthogonal complement; $\mathfrak{M}$ is termed the initial set of W, while the range of W is also termed its final set.

We denote complex or real numbers by a, a_i, b, b_i; integers will be represented by i, j, m, n, p, q. For a complex number a, the symbols $\mathfrak{R}a$ and $\mathfrak{J}a$ will stand for its real and imaginary part respectively. The abbreviations inf and sup will stand for the "greatest lower bound" and the "least upper bound" respectively.

We denote by α, β, ... crossnorms on the linear space $\mathfrak{B}_1 \odot \mathfrak{B}_2$ of expressions $\sum_{i=1}^{n} f_i \otimes g_i$. In particular, γ will stand for the greatest crossnorm, while λ represents the crossnorm furnished by the bound. α' is defined on $\mathfrak{B}_1^* \odot \mathfrak{B}_2^*$ and denotes the norm associated with α. The Cantor-Meray closure of the normed linear space $\mathfrak{B}_1 \odot_\alpha \mathfrak{B}_2$ is denoted by $\mathfrak{B}_1 \otimes_\alpha \mathfrak{B}_2$.

CHAPTER I

THE ALGEBRA OF EXPRESSIONS $\sum_{i=1}^{n} f_i \otimes g_i$.

1. The expressions $\sum_{i=1}^{n} f_i \otimes g_i$.

Throughout this chapter we assume that $\mathcal{R}_1$ and $\mathcal{R}_2$ denote any two linear spaces and $\mathcal{R}_1^*$, $\mathcal{R}_2^*$, denote the linear spaces of all additive functionals on $\mathcal{R}_1$ and $\mathcal{R}_2$ respectively.

We introduce two symbols $\otimes$ and $\cdot+\cdot$. With these for $f_1, \ldots\ldots, f_n$ in $\mathcal{R}_1$ and $g_1, \ldots\ldots, g_n$ in $\mathcal{R}_2$ we construct formal "expressions"

$$f_1 \otimes g_1 \cdot+\cdot f_2 \otimes g_2 \cdot+\cdot \ldots\ldots \cdot+\cdot f_n \otimes g_n .$$

We may abbreviate the last expression by writing $\sum_{i=1}^{n} f_i \otimes g_i$. Among these expressions we introduce a relation $\sim$ subject to the following rules:

(i) $f_1 \otimes g_1 \cdot+\cdot f_2 \otimes g_2 \cdot+\cdot \ldots\ldots \cdot+\cdot f_n \otimes g_n$

$$\sim f_{1'} \otimes g_{1'} \cdot+\cdot f_{2'} \otimes g_{2'} \cdot+\cdot \ldots\ldots \cdot+\cdot f_{n'} \otimes g_{n'}$$

where $1', 2', \ldots\ldots, n'$ denotes any permutation of the integers $1, 2, \ldots, n$.

(ii) $(f_1' + f_1'') \otimes g_1 \cdot+\cdot f_2 \otimes g_2 \cdot+\cdot \ldots\ldots \cdot+\cdot f_n \otimes g_n$

$$\sim f_1' \otimes g_1 \cdot+\cdot f_1'' \otimes g_1 \cdot+\cdot f_2 \otimes g_2 \cdot+\cdot \ldots\ldots \cdot+\cdot f_n \otimes g_n .$$

(ii') $f_1 \otimes (g_1' + g_1'') \cdot+\cdot f_2 \otimes g_2 \cdot+\cdot \ldots\ldots \cdot+\cdot f_n \otimes g_n$

$$\sim f_1 \otimes g_1' \cdot+\cdot f_1 \otimes g_1'' \cdot+\cdot f_2 \otimes g_2 \cdot+\cdot \ldots\ldots \cdot+\cdot f_n \otimes g_n .$$

(iii) $(a_1 f_1) \otimes g_1 \cdot+\cdot (a_2 f_2) \otimes g_2 \cdot+\cdot \ldots\ldots \cdot+\cdot (a_n f_n) \otimes g_n$

$$\sim f_1 \otimes (a_1 g_1) \cdot+\cdot f_2 \otimes (a_2 g_2) \cdot+\cdot \ldots\ldots \cdot+\cdot f_n \otimes (a_n g_n) .$$

DEFINITION 1.1. Two expressions $\sum_{i=1}^{n} f_i \otimes g_i$ and $\sum_{j=1}^{m} h_j \otimes k_j$ will be termed equivalent, if one can be transformed into the other by a finite number of successive applications of Rules (i), (ii), (ii'), (iii). We write this,

$$\sum_{i=1}^{n} f_i \otimes g_i \simeq \sum_{j=1}^{m} h_j \otimes k_j .$$

Rule (i) shows that $\simeq$ is reflexive, that is, every expression is equivalent to itself. The definition also implies transitivity.

Some elementary results can be readily obtained. For instance, if

$$\sum_{i=1}^{n} f_i \otimes g_i \simeq \sum_{j=1}^{m} h_j \otimes k_j \quad \text{and} \quad \sum_{i=1}^{n'} f_i' \otimes g_i' \simeq \sum_{j=1}^{m'} h_j' \otimes k_j'$$

then,

$$\sum_{i=1}^{n} f_i \otimes g_i \cdot + \cdot \sum_{i=1}^{n'} f_i' \otimes g_i' \simeq \sum_{j=1}^{m} h_j \otimes k_j \cdot + \cdot \sum_{j=1}^{m'} h_j' \otimes k_j' .$$

LEMMA 1.1. Every expression $\sum_{i=1}^{n} f_i \otimes g_i$ is equivalent to either $0 \otimes 0$ or to an expression $\sum_{j=1}^{m} h_j \otimes k_j$ in which both the $h_1, \ldots\ldots, h_m$ and $k_1, \ldots\ldots, k_m$ are linearly independent.

Proof. Suppose that in either set $f_1, \ldots\ldots, f_n$ or $g_1, \ldots\ldots, g_n$ the elements are linearly dependent. Then, $\sum_{i=1}^{n} f_i \otimes g_i$ is equivalent to an expression involving only $n-1$ terms. For instance, if $f_1 = \sum_{i=2}^{n} a_i f_i$ then

$$\begin{aligned} f_1 \otimes g_1 \cdot + \cdot \sum_{i=2}^{n} f_i \otimes g_i &\simeq \Big(\sum_{i=2}^{n} a_i f_i\Big) \otimes g_1 \cdot + \cdot \sum_{i=2}^{n} f_i \otimes g_i \\ &\simeq \sum_{i=2}^{n} (a_i f_i) \otimes g_1 \cdot + \cdot \sum_{i=2}^{n} f_i \otimes g_i \\ &\simeq \sum_{i=2}^{n} f_i \otimes (a_i g_1) \cdot + \cdot \sum_{i=2}^{n} f_i \otimes g_i \\ &\simeq \sum_{i=2}^{n} f_i \otimes (a_i g_1 + g_i) . \end{aligned}$$

We may therefore continue to reduce the number of terms until we have either $\sum_{j=1}^{m} h_j \otimes k_j$ in which both the $h_1, \ldots\ldots, h_m$ and $k_1, \ldots\ldots, k_m$ are linearly independent or $f \otimes 0$ or $0 \otimes g$. But $f \otimes 0 \simeq f \otimes (0.0) \simeq (0f) \otimes 0 \simeq 0 \otimes 0$. Similarly, $0 \otimes g \simeq 0 \otimes 0$.

2. The linear spaces $\mathcal{R}_1 \odot \mathcal{R}_2$ and $\mathcal{R}_1^* \odot \mathcal{R}_2^*$.

DEFINITION 1.2. We denote by $\mathcal{R}_1 \odot \mathcal{R}_2$ the set of all expressions of the form $\Sigma_{i=1}^{m} f_i \otimes g_i$. In the last set we consider equivalent expressions as identical, that is, the class of all expressions equivalent to each other we combine into a single element $\widetilde{f}$. If an expression is in $\widetilde{f}$ it will be termed "an expression for $\widetilde{f}$ ". Quite often we will find it convenient to permit "an expression for $\widetilde{f}$ " to stand for $\widetilde{f}$. If $\Sigma_{i=1}^{m} f_i \otimes g_i$ is an expression for $\widetilde{f}$, $\Sigma_{j=1}^{m} f_j' \otimes g_j'$ an expression for $\widetilde{g}$, then for a scalar a we define $a\widetilde{f}$ as the set of expressions equivalent to $\Sigma_{i=1}^{m} (af_i) \otimes g_i$, and $\widetilde{f} + \widetilde{g}$ as the set of expressions equivalent to $\Sigma_{i=1}^{m} f_i \otimes g_i \cdot + \cdot \Sigma_{j=1}^{m} f_j' \otimes g_j'$.

It is a consequence of Definition 1.1 and Rules (i), (ii), (ii'), (iii) that $a\widetilde{f}$ does not depend on the particular expression used. Similarly, $\widetilde{f} + \widetilde{g}$ is defined uniquely.

It is easy to see that the usual properties of addition and multiplication by a scalar hold; for instance,

$$\widetilde{f} + \widetilde{g} = \widetilde{g} + \widetilde{f} ; \qquad \widetilde{f} + (\widetilde{g} + \widetilde{h}) = (\widetilde{f} + \widetilde{g}) + \widetilde{h} ,$$

$$a(\widetilde{f} + \widetilde{g}) = a\widetilde{f} + a\widetilde{g} ; \qquad a(b\widetilde{f}) = (ab)\widetilde{f} .$$

The zero element $\widetilde{0}$ is the class of all expressions equivalent to $0 \otimes 0$.

THEOREM 1.1. $\mathcal{R}_1 \odot \mathcal{R}_2$ is a linear set, that is, a commutative group with scalar operators.

Proof. The proof follows from the preceding discussion.

Similarly, we form the linear set $\mathcal{R}_1^* \otimes \mathcal{R}_2^*$ of expressions $\sum_{j=1}^{m} F_j \otimes G_j$, (where $F_1, \ldots\ldots, F_m$ are in $\mathcal{R}_1^*$ while $G_1, \ldots\ldots, G_m$ are in $\mathcal{R}_2^*$).

A little more about these expressions, in particular the invariance of their "rank" under equivalence, may be found in [12]. In the present chapter we only state those properties of expressions which will be needed in our future discussions.

3. Transformations on expressions.

LEMMA 1.2. Let $F \in \mathcal{R}_1^*$. Then, $\sum_{i=1}^{n} f_i \otimes g_i \simeq \sum_{j=1}^{m} h_j \otimes k_j$ implies $\sum_{i=1}^{n} F(f_i) g_i = \sum_{j=1}^{m} F(h_j) k_j$.

Proof. For a given expression $\sum_{i=1}^{n} f_i \otimes g_i$ we form the transformation $T(\sum_{i=1}^{n} f_i \otimes g_i) = \sum_{i=1}^{n} F(f_i) g_i$. Since $F \in \mathcal{R}_1^*$, the values of T remain the same for both sides of any of the $\sim$ relationships expressed by Rules (i)--(iii). This means that a single application of these Rules to an expression does not change the value of T. Thus, a finite number of successive applications of Rules (i)--(iii) to an expression does not change the value of T for that expression. This concludes the proof.

Similarly, we prove the following "dual" of the last statement:

LEMMA 1.3. Let $f \in \mathcal{R}_1$. Then, $\sum_{i=1}^{n} F_i \otimes G_i \simeq \sum_{j=1}^{m} H_j \otimes K_j$ implies $\sum_{i=1}^{n} F_i(f) G_i = \sum_{j=1}^{m} H_j(f) K_j$.

DEFINITION 1.3. Let $\sum_{j=1}^{m} F_j \otimes G_j$ be an expression in $\mathcal{R}_1^* \otimes \mathcal{R}_2^*$ while $\sum_{i=1}^{n} f_i \otimes g_i$ represents an expression in $\mathcal{R}_1 \otimes \mathcal{R}_2$. Under their "inner product" in symbol $(\sum_{j=1}^{m} F_j \otimes G_j)(\sum_{i=1}^{n} f_i \otimes g_i)$ we understand the number $\sum_{j=1}^{m} \sum_{i=1}^{n} F_j(f_i) G_j(g_i)$.

LEMMA 1.4. The inner product is invariant under equivalence. We mean hereby, that

$$\sum_{i=1}^{n} f_i \otimes g_i \simeq \sum_{i=1}^{r} h_i \otimes k_i \quad \text{and} \quad \sum_{j=1}^{m} F_j \otimes G_j \simeq \sum_{j=1}^{s} H_j \otimes K_j$$

imply

$$(\sum_{j=1}^{m} F_j \otimes G_j)(\sum_{i=1}^{n} f_i \otimes g_i) = (\sum_{j=1}^{s} H_j \otimes K_j)(\sum_{i=1}^{r} h_i \otimes k_i) .$$

Proof. Since $\sum_{j=1}^{m} F_j \otimes G_j \simeq \sum_{j=1}^{s} H_j \otimes K_j$, Lemma 1.3 gives,

$$(\sum_{j=1}^{m} F_j \otimes G_j)(\sum_{i=1}^{n} f_i \otimes g_i) = (\sum_{j=1}^{s} H_j \otimes K_j)(\sum_{i=1}^{n} f_i \otimes g_i) .$$

Similarly, by Lemma 1.2,

$$(\sum_{j=1}^{s} H_j \otimes K_j)(\sum_{i=1}^{n} f_i \otimes g_i) = (\sum_{j=1}^{s} H_j \otimes K_j)(\sum_{i=1}^{r} h_i \otimes k_i) .$$

The last two equalities furnish the desired proof.

The proof of the following two lemmas is obtained analogously:

LEMMA 1.5. Let A denote an additive transformation from $\mathfrak{R}_1$ into $\mathfrak{R}_2^*$. Then,

$$\sum_{i=1}^{n} f_i \otimes g_i \simeq \sum_{j=1}^{m} h_j \otimes k_j \quad \text{implies} \quad \sum_{i=1}^{n} (Af_i) g_i = \sum_{j=1}^{m} (Ah_j) k_j .$$

LEMMA 1.6. Let S and T denote two additive transformations on $\mathfrak{R}_1$ and $\mathfrak{R}_2$ respectively. Then,

$$\sum_{i=1}^{n} f_i \otimes g_i \simeq \sum_{j=1}^{m} h_j \otimes k_j \quad \text{implies} \quad \sum_{i=1}^{n} Sf_i \otimes Tg_i \simeq \sum_{j=1}^{m} Sh_j \otimes Tk_j .$$

DEFINITION 1.4. For a fixed expression $\sum_{i=1}^{n} f_i \otimes g_i$ in $\mathfrak{R}_1 \odot \mathfrak{R}_2$ we define the following transformation from $\mathfrak{R}_1^*$ into $\mathfrak{R}_2$:

$$T_{\sum_{i=1}^{n} f_i \otimes g_i}(F) = \sum_{i=1}^{n} F(f_i) g_i \qquad \text{for } F \in \mathfrak{R}_1^* .$$

It is a consequence of Lemma 1.2 that equivalent expressions furnish the same transformation of Definition 1.4.

LEMMA 1.7. $T_{\sum_{i=1}^{m} f_i \otimes g_i} = 0$ implies $\sum_{i=1}^{m} f_i \otimes g_i \simeq 0 \otimes 0$.

Proof. Suppose $\sum_{i=1}^{m} f_i \otimes g_i$ is not equivalent to $0 \otimes 0$. Then by Lemma 1.1, $\sum_{i=1}^{m} f_i \otimes g_i \simeq \sum_{j=1}^{m} h_j \otimes k_j$ where the h_j's as well as the k_j's are linearly independent. In particular, $h_1 \neq 0$. Consequently we can find an $F \in \mathcal{B}_1^*$ for which $F(h_1) \neq 0$. The linear independence of the k_j's implies $\sum_{j=1}^{m} F(h_j)k_j \neq 0$. An application of Lemma 1.2 furnishes $\sum_{i=1}^{m} F(f_i)g_i \neq 0$ and therefore,

$$T_{\sum_{i=1}^{m} f_i \otimes g_i}(F) \neq 0 \quad .$$

This concludes the proof.

LEMMA 1.8. Two expressions $\sum_{i=1}^{m} f_i \otimes g_i$ and $\sum_{j=1}^{m} h_j \otimes k_j$ are equivalent if and only if, they furnish the same transformation of Definition 1.4.

Proof. The proof is a consequence of Lemmas 1.2 and 1.7.

CHAPTER II

CROSSNORMS

1. The normed linear spaces $\mathfrak{B}_1 \odot_\alpha \mathfrak{B}_2$ and $\mathfrak{B}_1^* \odot_{\alpha'} \mathfrak{B}_2^*$.

Henceforth, we assume that $\mathfrak{B}_1$ and $\mathfrak{B}_2$ denote any two Banach spaces while $\mathfrak{B}_1^*$ and $\mathfrak{B}_2^*$ stand for their conjugate spaces, that is, the spaces of all additive and bounded functionals on $\mathfrak{B}_1$ and $\mathfrak{B}_2$ respectively.

Clearly, an expression $\sum_{i=1}^{n} f_i \otimes g_i$ in $\mathfrak{B}_1 \odot \mathfrak{B}_2$ may be interpreted as an operator A of finite rank from $\mathfrak{B}_1^*$ into $\mathfrak{B}_2$ whose defining equation is given by

$$A(F) = \sum_{i=1}^{n} F(f_i) g_i \qquad \text{for } F \in \mathfrak{B}_1^*.$$

By Lemma 1.8, two expressions are equivalent if and only if, they furnish the same operator of finite rank. In the case when $\mathfrak{B}_1$ is reflexive, then also every operator A of finite rank from $\mathfrak{B}_1^*$ into $\mathfrak{B}_2$ is determined by many such expressions. To see this, suppose that the range of A is determined by the linearly independent $g_1, \ldots\ldots, g_n$. We have, $A(F) = \sum_{i=1}^{n} a_i(F) g_i$ where $a_i(F)$ represents additive bounded functionals on $\mathfrak{B}_1^*$. Since $\mathfrak{B}_1$ is assumed to be reflexive, there exist elements $f_1, \ldots\ldots, f_n$ in $\mathfrak{B}_1$ such that $a_i(F) = F(f_i)$ for $F \in \mathfrak{B}_1^*$. Thus, A is determined by $\sum_{i=1}^{n} f_i \otimes g_i$. We may sum this up as follows: For any two Banach spaces $\mathfrak{B}_1$ and $\mathfrak{B}_2$, the linear space $\mathfrak{B}_1 \odot \mathfrak{B}_2$ may be considered as a subspace of the space of all operators from $\mathfrak{B}_1^*$ into $\mathfrak{B}_2$ of finite rank. In case $\mathfrak{B}_1$ is assumed to be reflexive, then $\mathfrak{B}_1 \odot \mathfrak{B}_2$ represents precisely the class of all operators from $\mathfrak{B}_1^*$ into $\mathfrak{B}_2$ of finite rank.

Throughout this and the following Chapters III and IV we shall assume that $\mathfrak{B}_1$ and $\mathfrak{B}_2$ represents perfectly general Banach spaces without any special restrictions.

DEFINITION 2.1. Under a norm α in $\mathfrak{B}_1 \odot \mathfrak{B}_2$ we shall understand any non-negative function of expressions $\sum_{i=1}^n f_i \otimes g_i$ satisfying the following conditions:

I. $\alpha(\sum_{i=1}^n f_i \otimes g_i) = 0$ if and only if, $\sum_{i=1}^n f_i \otimes g_i \simeq 0 \otimes 0$.

II. $\alpha(a \sum_{i=1}^n f_i \otimes g_i) = |a| \alpha(\sum_{i=1}^n f_i \otimes g_i)$ for any constant a .

III. $\alpha(\sum_{i=1}^n f_i \otimes g_i \cdot + \cdot \sum_{j=1}^m f_j' \otimes g_j') \leq \alpha(\sum_{i=1}^n f_i \otimes g_i) + \alpha(\sum_{j=1}^m f_j' \otimes g_j')$.

IV. $\alpha(\sum_{i=1}^n f_i \otimes g_i) = \alpha(\sum_{j=1}^m f_j' \otimes g_j')$ if $\sum_{i=1}^n f_i \otimes g_i \simeq \sum_{j=1}^m f_j' \otimes g_j'$.

DEFINITION 2.2. Let α denote a given norm on $\mathfrak{B}_1 \odot \mathfrak{B}_2$. For a fixed expression $\sum_{j=1}^m F_j \otimes G_j$ in $\mathfrak{B}_1^* \odot \mathfrak{B}_2^*$ we define $\alpha'(\sum_{j=1}^m F_j \otimes G_j)$ as the least (finite or infinite) constant C satisfying the inequality:

$$|(\sum_{j=1}^m F_j \otimes G_j)(\sum_{i=1}^n f_i \otimes g_i)| \leq C\, \alpha(\sum_{i=1}^n f_i \otimes g_i)$$

for all expressions $\sum_{i=1}^n f_i \otimes g_i$ in $\mathfrak{B}_1 \odot \mathfrak{B}_2$. Thus,

$$\alpha'(\sum_{j=1}^m F_j \otimes G_j) = \sup_{\sum_{i=1}^n f_i \otimes g_i} \frac{|(\sum_{j=1}^m F_j \otimes G_j)(\sum_{i=1}^n f_i \otimes g_i)|}{\alpha(\sum_{i=1}^n f_i \otimes g_i)} .$$

$\alpha'(\sum_{j=1}^m F_j \otimes G_j)$ is therefore a function of expressions in $\mathfrak{B}_1^* \odot \mathfrak{B}_2^*$.

LEMMA 2.1. Whenever α' is finite for every expression in $\mathfrak{B}_1^* \odot \mathfrak{B}_2^*$, then it also satisfies conditions I--IV; α' is termed the norm "associated" with α .

Proof. Let $\sum_{j=1}^{m} F_j \otimes G_j \simeq 0 \otimes 0$. By Lemma 1.4, we have $(\sum_{j=1}^{m} F_j \otimes G_j)(\sum_{i=1}^{n} f_i \otimes g_i) = 0$ for every $\sum_{i=1}^{n} f_i \otimes g_i$. Thus, $\alpha'(\sum_{j=1}^{m} F_j \otimes G_j) = 0$. Suppose on the other hand that $\sum_{j=1}^{m} F_j \otimes G_j$ is not equivalent to $0 \otimes 0$. By Lemma 1.1, $\sum_{j=1}^{m} F_j \otimes G_j \simeq \sum_{i=1}^{s} F_i' \otimes G_i'$ with both the $F_1', \ldots\ldots, F_s'$ as well as the $G_1', \ldots\ldots, G_s'$ linearly independent. Since, $F_1' \neq 0$, we can choose an $f \in \mathcal{B}_1$ with $F_1'(f) \neq 0$. The linear independence of the G_i''s implies $\sum_{i=1}^{s} F_i'(f) G_i' \neq 0$. Now choose a $g \in \mathcal{B}_2$ such that $\sum_{i=1}^{s} F_i'(f) G_i'(g) \neq 0$. By Lemma 1.3, also $\sum_{j=1}^{m} F_j(f) G_j(g) \neq 0$. Since, obviously $\alpha(f \otimes g) > 0$, we must have $\alpha'(\sum_{j=1}^{m} F_j \otimes G_j) > 0$.

II and III are immediate.

IV is a consequence of Lemma 1.4 and the definition of α' for a given α. This concludes the proof.

DEFINITION 2.3. The linear set $\mathcal{B}_1 \odot \mathcal{B}_2$ (Definition 1.2) on which we define a norm α (Definition 2.1) will be denoted by $\mathcal{B}_1 \odot_\alpha \mathcal{B}_2$. Similarly, we define $\mathcal{B}_1^* \odot_{\alpha'} \mathcal{B}_2^*$.

LEMMA 2.2. Let α and β represent two norms on $\mathcal{B}_1 \odot \mathcal{B}_2$. Then, $\alpha \leq \beta$ implies $\alpha' \geq \beta'$ on $\mathcal{B}_1^* \odot \mathcal{B}_2^*$.

Proof. The proof is a consequence of Definition 2.2.

2. Crossnorms.

Among the norms α on $\mathcal{B}_1 \odot \mathcal{B}_2$ of particular interest are the "crossnorms", that is, norms satisfying the following additional condition:

V. $\alpha(f \otimes g) = \|f\| \, \|g\|$.

Therefore, a norm is a crossnorm if and only if, its value for the expressions which generate operators from $\mathcal{B}_1^*$ into $\mathcal{B}_2$ of rank ≤ 1 is equal to the bound of the generated operators.

At this point it seems proper to mention the following fact which we shall make use of in the future: In case both $\mathcal{B}_1$ and $\mathcal{B}_2$ are separable, then for a given crossnorm α , $\mathcal{B}_1 \odot_\alpha \mathcal{B}_2$ is also separable. This is expressed in detail in the following Lemmas 2.3 and 2.4.

LEMMA 2.3. A crossnorm $\alpha(\sum_{i=1}^n f_i \otimes g_i)$ is a continuous function of the f_i's and the g_i's , that is, for a given $\varepsilon > 0$ we can find a $\delta = \delta(f_1, \ldots, f_n;\ g_1, \ldots, g_n) > 0$ such that $\|f_i - f_i'\| < \delta$, $\|g_i - g_i'\| < \delta$, for $i = 1, 2, \ldots, n$ implies

$$\alpha(\textstyle\sum_{i=1}^n f_i \otimes g_i \cdot - \cdot \sum_{i=1}^n f_i' \otimes g_i') < \varepsilon .$$

Proof. We verify without difficulty the following relation:

$$\textstyle\sum_{i=1}^n f_i \otimes g_i \cdot - \cdot \sum_{i=1}^n f_i' \otimes g_i' \simeq$$

$$\textstyle\sum_{i=1}^n (f_i - f_i') \otimes g_i \cdot + \cdot \sum_{i=1}^n f_i \otimes (g_i - g_i') \cdot + \cdot \sum_{i=1}^n (f_i - f_i') \otimes (g_i - g_i') .$$

Therefore,

$$\alpha(\textstyle\sum_{i=1}^n f_i \otimes g_i \cdot - \cdot \sum_{i=1}^n f_i' \otimes g_i') \leq \alpha(\sum_{i=1}^n (f_i - f_i') \otimes g_i) +$$

$$\alpha(\textstyle\sum_{i=1}^n f_i \otimes (g_i - g_i')) + \alpha(\sum_{i=1}^n (f_i - f_i') \otimes (g_i - g_i')) \leq$$

$$\textstyle\sum_{i=1}^n \|f_i - f_i'\| \, \|g_i\| + \sum_{i=1}^n \|f_i\| \, \|g_i - g_i'\| + \sum_{i=1}^n \|f_i - f_i'\| \, \|g_i - g_i'\| .$$

This concludes the proof.

LEMMA 2.4. Let $\mathcal{B}_1$ and $\mathcal{B}_2$ be separable. Then for any crossnorm α , the normed linear space $\mathcal{B}_1 \odot_\alpha \mathcal{B}_2$ is also separable.

Proof. Let $f_1, f_2, \ldots$ and $g_1, g_2, \ldots$ denote two sequences of elements dense in $\mathfrak{B}_1$ and $\mathfrak{B}_2$ respectively. Then the set of expressions

$$\Sigma_{i=1}^{n} f_{t_i} \otimes g_{t'_i} \quad ; \quad t_i, t'_i = 1,2,\ldots, \quad n=1,2,\ldots$$

is denumerable and dense in $\mathfrak{B}_1 \odot_\alpha \mathfrak{B}_2$. Therefore, the set of elements for which these expressions stand is dense in $\mathfrak{B}_1 \odot_\alpha \mathfrak{B}_2$.

LEMMA 2.5. For any crossnorm α on $\mathfrak{B}_1 \odot \mathfrak{B}_2$,

$$\alpha'(F \otimes G) \geq \|F\| \, \|G\| \qquad \text{for } F \in \mathfrak{B}_1^*, \; G \in \mathfrak{B}_2^*.$$

Proof. Let F and G be fixed. It is a consequence of Definition 2.2 that for any pair f, g we have,

$$|F(f)G(g)| \leq \alpha'(F \otimes G) \; \alpha(f \otimes g) = \alpha'(F \otimes G) \|f\| \, \|g\| .$$

This clearly implies $\alpha'(F \otimes G) \geq \|F\| \, \|G\|$. This concludes the proof.

REMARK 2.1. Let α be a given norm on $\mathfrak{B}_1 \odot \mathfrak{B}_2$. Whenever $\alpha'(F \otimes G)$ is finite for every pair $F \in \mathfrak{B}_1^*$, $G \in \mathfrak{B}_2^*$, then $\alpha'(\Sigma_{j=1}^{m} F_j \otimes G_j) \leq \Sigma_{j=1}^{m} \alpha'(F_j \otimes G_j)$ is also finite for every expression in $\mathfrak{B}_1^* \odot \mathfrak{B}_2^*$. It follows then that α' is also a norm.

Quite often we shall assume that the crossnorm α is uniform, that is, satisfies the following condition:

VI. $\alpha(\Sigma_{i=1}^{n} Sf_i \otimes Tg_i) \leq |||S||| \, |||T||| \; \alpha(\Sigma_{i=1}^{n} f_i \otimes g_i)$

for every pair of operators S and T on $\mathfrak{B}_1$ and $\mathfrak{B}_2$ respectively.

The geometric significance of this condition is clear. Furthermore, for S and T as above, we define an operator $S \otimes T$ on $\mathfrak{B}_1 \odot_\alpha \mathfrak{B}_2$ as follows:

$$(S \otimes T)(\Sigma_{i=1}^{n} f_i \otimes g_i) = \Sigma_{i=1}^{n} Sf_i \otimes Tg_i .$$

This operator is uniquely defined, since it is invariant under equivalence. We mean hereby $\sum_{i=1}^{m} f_i \otimes g_i \simeq \sum_{j=1}^{m} h_j \otimes k_j$ implies $\sum_{i=1}^{m} Sf_i \otimes Tg_i \simeq \sum_{j=1}^{m} Sh_j \otimes Tk_j$ (Lemma 1.6). It is readily seen that a crossnorm α on $\mathfrak{B}_1 \odot \mathfrak{B}_2$ is uniform if, and only if, the operator $S \otimes T$ on $\mathfrak{B}_1 \odot_\alpha \mathfrak{B}_2$ satisfies the condition

$$|||S \otimes T||| = |||S||| \; |||T||| \; .$$

We shall see later that the interesting crossnorms are those which satisfy condition VI.

3. The bound as a crossnorm.

DEFINITION 2.4. For $\sum_{i=1}^{m} f_i \otimes g_i$ in $\mathfrak{B}_1 \odot \mathfrak{B}_2$ we define

$$\lambda\left(\sum_{i=1}^{m} f_i \otimes g_i\right) = \sup_{F,G} \frac{\left|\sum_{i=1}^{m} F(f_i)G(g_i)\right|}{\|F\| \; \|G\|}$$

where sup is taken over the set of all numbers obtained when $0 \neq F$ and $0 \neq G$ varies in $\mathfrak{B}_1^*$ and $\mathfrak{B}_2^*$ respectively.

Clearly, $\lambda(\sum_{i=1}^{m} f_i \otimes g_i)$ represents the bound of the operator $\sum_{i=1}^{m} F(f_i) g_i$ from $\mathfrak{B}_1^*$ into $\mathfrak{B}_2$ determined by the expression $\sum_{i=1}^{m} f_i \otimes g_i$.

LEMMA 2.6. λ is a uniform crossnorm.

Proof. Definition 2.4 gives $\lambda(\sum_{i=1}^{m} f_i \otimes g_i) = 0$ if and only if, $\sum_{i=1}^{m} F(f_i)G(g_i) = 0$ for all $F \in \mathfrak{B}_1^*$, $G \in \mathfrak{B}_2^*$. The last happens if and only if, $\sum_{i=1}^{m} F(f_i) g_i = 0$ for all $F \in \mathfrak{B}_1^*$, and therefore if and only if, $\sum_{i=1}^{m} f_i \otimes g_i \simeq 0 \otimes 0$ as may be concluded from Lemma 1.8. Thus, λ satisfies condition I.

That λ satisfies conditions II and III is immediate.

Condition IV is a consequence of Lemma 1.4.

We shall check condition V:

$$\lambda(f \otimes g) = \sup_{F,G} \frac{|F(f)G(g)|}{\|F\| \, \|G\|}$$

$$\sup_F \frac{|F(f)|}{\|F\|} \quad \sup_G \frac{|G(g)|}{\|G\|} = \|f\| \, \|g\| .$$

Finally we shall prove that λ satisfies condition VI. Let S and T denote two operators on $\mathfrak{B}_1$ and $\mathfrak{B}_2$ respectively. Their adjoints S^* and T^* represent operators on $\mathfrak{B}_1^*$ and $\mathfrak{B}_2^*$ respectively. Clearly,

$$\|S^*(F)\| \leq |||S^*||| \, \|F\| = |||S||| \, \|F\| \quad \text{and}$$

$$\|T^*(G)\| \leq |||T^*||| \, \|G\| = |||T||| \, \|G\| .$$

Definition 2.2 gives

$$|(F \otimes G)(\Sigma_{i=1}^{m} Sf_i \otimes Tg_i)| = |(S^*F \otimes T^*G)(\Sigma_{i=1}^{m} f_i \otimes g_i)|$$

$$\leq \lambda'(S^*F \otimes T^*G) \; \lambda(\Sigma_{i=1}^{m} f_i \otimes g_i) .$$

The last -- since λ' is a crossnorm (as is proven in Lemma 2.7 which follows) -- is clearly

$$= \|S^*F\| \, \|T^*G\| \quad \lambda(\Sigma_{i=1}^{m} f_i \otimes g_i)$$

$$\leq \|F\| \, \|G\| \, |||S||| \, |||T||| \; \lambda(\Sigma_{i=1}^{m} f_i \otimes g_i) .$$

Since our inequality holds for any pair F, G, Definition 2.4 gives

$$\lambda(\Sigma_{i=1}^{m} Sf_i \otimes Tg_i) \leq |||S||| \, |||T||| \quad \lambda(\Sigma_{i=1}^{m} f_i \otimes g_i) .$$

This concludes the proof.

LEMMA 2.7. λ' is a crossnorm.

Proof. Let F and G be fixed. It is a consequence of Definition 2.4 that for any expression $\Sigma_{i=1}^{m} f_i \otimes g_i$ we have,

$$|(F \otimes G)(\sum_{i=1}^{m} f_i \otimes g_i)| \leq \|F\| \|G\| \lambda(\sum_{i=1}^{m} f_i \otimes g_i) .$$

Thus, Definition 2.2 gives $\lambda'(F \otimes G) \leq \|F\| \|G\|$. This together with Lemma 2.5 proves that $\lambda'(F \otimes G) = \|F\| \|G\|$. By Remark 2.1, λ' is finite for every expression $\sum_{j=1}^{m} F_j \otimes G_j$. Consequently, λ' is a crossnorm by Lemma 2.1. This concludes the proof.

LEMMA 2.8. The associate α' with a crossnorm $\alpha \geqq \lambda$ is also a crossnorm.

Proof. Lemma 2.2 gives $\alpha' \leq \lambda'$. In particular, Lemma 2.7 furnishes $\alpha'(F \otimes G) \leq \lambda'(F \otimes G) = \|F\| \|G\|$. An application of Lemma 2.5 concludes the proof.

THEOREM 2.1. The associate α' with a crossnorm α is also a crossnorm if and only if, $\alpha \geqq \lambda$. Therefore, λ represents the least crossnorm whose associate is also a crossnorm.

Proof. Suppose that for a crossnorm α and an expression $\sum_{i=1}^{m} f_i \otimes g_i$ we have

$$\alpha(\sum_{i=1}^{m} f_i \otimes g_i) < \lambda(\sum_{i=1}^{m} f_i \otimes g_i) .$$

By Definition 2.4, there exists an $F \in \mathfrak{B}_1^*$ and a $G \in \mathfrak{B}_2^*$ such that

$$\alpha(\sum_{i=1}^{m} f_i \otimes g_i) \|F\| \|G\| < |(F \otimes G)(\sum_{i=1}^{m} f_i \otimes g_i)| .$$

By Definition 2.2, the right side of the last inequality is not greater than $\alpha'(F \otimes G)\, \alpha(\sum_{i=1}^{m} f_i \otimes g_i)$. Thus, $\alpha'(F \otimes G) > \|F\| \|G\|$, that is, α' is not a crossnorm. Therefore, whenever for a crossnorm α its associate α' is also a crossnorm, then $\alpha \geqq \lambda$. The converse was proven in Lemma 2.8. This concludes the proof.

REMARK 2.2. At this point it seems proper to emphasize the "general character" of the crossnorm λ. We mean hereby that for any two Banach spaces $\mathfrak{B}_1$ and $\mathfrak{B}_2$ without any special restrictions, the crossnorm λ on $\mathfrak{B}_1 \odot \mathfrak{B}_2$ can be always constructed.

LEMMA 2.9. Let $F_i \in \mathfrak{B}_1^*$, $G_i \in \mathfrak{B}_2^*$, for $i = 1, \ldots\ldots, n$. Then,

(1) $\sup |\sum_{i=1}^{n} F_i(f) G_i(g)|$ for all $f \in \mathfrak{B}_1$, $g \in \mathfrak{B}_2$; $\|f\| = \|g\| = 1$

is equal to

(2) $\sup |\sum_{i=1}^{n} \mathfrak{F}(F_i) \mathfrak{G}(G_i)|$ for all $\mathfrak{F} \in \mathfrak{B}_1^{**}$, $\mathfrak{G} \in \mathfrak{B}_2^{**}$; $\|\mathfrak{F}\| = \|\mathfrak{G}\| = 1$.

Proof. We remark first that if F_0 represents an additive bounded functional on a Banach space $\mathfrak{B}$ then

$$\sup |F_0(f)| \qquad \text{for all} \quad f \in \mathfrak{B},\ \|f\| = 1$$

is equal to

$$\sup |\mathfrak{F}(F_0)| \qquad \text{for all} \quad \mathfrak{F} \in \mathfrak{B}^{**},\ \|\mathfrak{F}\| = 1 .$$

Both numbers obviously represent $\|F_0\|$ [1, pp. 54-55]. Therefore for a fixed $g_0 \in \mathfrak{B}_2$, $\|g_0\| = 1$, substituting $G_i(g_0)F_i$ for F_0 we get

$$\sup |\sum_{i=1}^{n} F_i(f) G_i(g_0)| \qquad \text{for} \quad f \in \mathfrak{B}_1,\ \|f\| = 1$$

is equal to

$$\sup |\sum_{i=1}^{n} \mathfrak{F}(F_i) G_i(g_0)| \qquad \text{for} \quad \mathfrak{F} \in \mathfrak{B}_1^{**},\ \|\mathfrak{F}\| = 1 .$$

This proves that (1) equals to

$$\sup |\sum_{i=1}^{n} \mathfrak{F}(F_i) G_i(g)| \qquad \text{for all } \mathfrak{F} \in \mathfrak{B}_1^{**},\ g \in \mathfrak{B}_2;\ \|\mathfrak{F}\| = \|g\| = 1 .$$

A similar reasoning proves that the last number equals to (2). This concludes the proof.

The preceding Lemma permits to express λ in a different form, for

Banach spaces which are conjugates of other Banach spaces. This forms the content of the following Lemma:

LEMMA 2.10. For an expression $\Sigma_{j=1}^{m} F_j \otimes G_j$ in $\mathcal{B}_1^* \odot \mathcal{B}_2^*$, $\lambda(\Sigma_{j=1}^{m} F_j \otimes G_j)$ may be also represented as

$$\sup_{f,g} \frac{|\Sigma_{j=1}^{m} F_j(f) G_j(g)|}{\|f\| \, \|g\|} .$$

Proof. This is a consequence of Definition 2.3 and Lemma 2.9.

LEMMA 2.11. Let α represent a crossnorm $\geqq \lambda$ on $\mathcal{B}_1 \odot \mathcal{B}_2$. Then α' is also a crossnorm $\geqq \lambda$ on $\mathcal{B}_1^* \odot \mathcal{B}_2^*$.

Proof. That α' is a crossnorm is stated in Theorem 2.1. Definition 2.2 gives

$$\alpha'(\Sigma_{j=1}^{m} F_j \otimes G_j) \geq \sup_{f,g} \frac{|(\Sigma_{j=1}^{m} F_j \otimes G_j)(f \otimes g)|}{\alpha(f \otimes g)}$$

$$= \sup_{f,g} \frac{|(\Sigma_{j=1}^{m} F_j \otimes G_j)(f \otimes g)|}{\|f\| \, \|g\|}$$

By Lemma 2.10 the last number equals to $\lambda(\Sigma_{j=1}^{m} F_j \otimes G_j)$. This concludes the proof.

THEOREM 2.2. Let α denote a crossnorm whose associate α' is also a crossnorm. Then, α'', α''', ... are also crossnorms.

Proof. Since α and α' are crossnorms, we have $\alpha \geqq \lambda$ by Theorem 2.1. Applying successively Lemma 2.11 we get $\alpha' \geqq \lambda$ on $\mathcal{B}_1^* \odot \mathcal{B}_2^*$, $\alpha'' \geqq \lambda$ on $\mathcal{B}_1^{**} \odot \mathcal{B}_2^{**}$, ... Thus, by Theorem 2.1, α'', α''', ... are crossnorms. This concludes the proof.

Theorem 2.1 states that λ represents the least crossnorms whose associate is also a crossnorm. We remark that λ does not necessarily represent the least crossnorm. In fact, we shall prove later (Chapter V, § 11) that when $\mathfrak{B}_1$ and $\mathfrak{B}_2$ denote two two-dimensional Euclidean spaces a least crossnorm in $\mathfrak{B}_1 \odot \mathfrak{B}_2$ does not exist. By Theorem 2.1 this implies the existence of crossnorms whose associates are not crossnorms.

The following Lemma concerning the "local character" of λ is of interest. It proves that $\lambda(\sum_{i=1}^{m} f_i \otimes g_i)$ depends only on the spatial relations between the f_i's and those between the g_i's and not on the including them spaces.

LEMMA 2.12. Let $\mathfrak{M}_1$ and $\mathfrak{M}_2$ denote two closed linear manifolds in the Banach spaces $\mathfrak{B}_1$ and $\mathfrak{B}_2$ respectively. Then λ on $\mathfrak{B}_1 \odot \mathfrak{B}_2$ is an extension of λ on $\mathfrak{M}_1 \odot \mathfrak{M}_2$.

Proof. Let $\sum_{i=1}^{m} f_i \otimes g_i$ be an expression in $\mathfrak{M}_1 \odot \mathfrak{M}_2 \subset \mathfrak{B}_1 \odot \mathfrak{B}_2$. We shall prove our Lemma, by showing the equality of the following two numbers (Definition 2.4):

(a) $\sup |\sum_{i=1}^{m} F(f_i)G(g_i)|$ for $F \in \mathfrak{M}_1^*$, $G \in \mathfrak{M}_2^*$; $\|F\| = \|G\| = 1$

and

(b) $\sup |\sum_{i=1}^{m} F(f_i)G(g_i)|$ for $F \in \mathfrak{B}_1^*$, $G \in \mathfrak{B}_2^*$; $\|F\| = \|G\| = 1$.

We recall Hahn-Banach's extension theorem [1, p. 55] which states that an additive and bounded functional F on a linear manifold $\mathfrak{M}$ in a Banach space $\mathfrak{B}$ can be extended (quite often in a non-unique manner) to an additive and bounded functional F° on $\mathfrak{B}$, without affecting the value of the bound, that is,

$$F^{\circ}(f) = F(f) \quad \text{for} \quad f \in \mathfrak{M} \qquad \text{and}$$

$$\sup_{0 \neq f \in \mathfrak{B}} \frac{|F^{\circ}(f)|}{\|f\|} = \sup_{0 \neq f \in \mathfrak{M}} \frac{|F(f)|}{\|f\|} .$$

From this theorem it is obvious that (a) $\leq$ (b) .

On the other hand, when F° denotes a given additive and bounded functional on $\mathfrak{B}_1$, we may restrict ourselves and consider the values of this functional only on the closed linear manifold $\mathfrak{M}_1 \subset \mathfrak{B}_1$. Thus, F° on $\mathfrak{B}_1$ defines an additive bounded functional F_0 on $\mathfrak{M}_1$ (such that $F^{\circ}(f) = F_0(f)$ for $f \in \mathfrak{M}_1$). It is clear that $\|F^{\circ}\| \geq \|F_0\|$ [1, p. 54] . Now, let F° and G° denote two non-zero additive bounded functionals on $\mathfrak{B}_1$ and $\mathfrak{B}_2$ respectively, while F_0 and G_0 stand for the corresponding functionals on $\mathfrak{M}_1$ and $\mathfrak{M}_2$ respectively. Then,

$$\frac{|(F^{\circ} \otimes G^{\circ})(\sum_{i=1}^{m} f_i \otimes g_i)|}{\|F^{\circ}\| \, \|G^{\circ}\|} \leq \frac{|(F_0 \otimes G_0)(\sum_{i=1}^{m} f_i \otimes g_i)|}{\|F_0\| \, \|G_0\|}$$

The last inequality holds for any $0 \neq F^{\circ} \in \mathfrak{B}_1^*$, $0 \neq G^{\circ} \in \mathfrak{B}_2^*$. Therefore, (b) $\leq$ (a) . This concludes the proof.

4. The greatest crossnorm.

In the arguments which follow we prove the existence and actually construct the unique greatest crossnorm.

DEFINITION 2.5. Let $\sum_{i=1}^{m} f_i^{\circ} \otimes g_i^{\circ}$ be a fixed expression in $\mathfrak{B}_1 \odot \mathfrak{B}_2$. We define

$$\gamma\left(\sum_{i=1}^{m} f_i^{\circ} \otimes g_i^{\circ}\right) = \inf \sum_{j=1}^{m} \|f_j\| \, \|g_j\|$$

where inf is taken over the set of all expressions $\sum_{j=1}^{m} f_j \otimes g_j$ equivalent to $\sum_{i=1}^{m} f_i^0 \otimes g_i^0$.

LEMMA 2.13. γ is a uniform crossnorm, that is, γ satisfies conditions I--VI.

Proof. I: Let $\sum_{i=1}^{m} f_i \otimes g_i$ denote a fixed expression. For an $F \in \mathcal{B}_1^*$ with $\|F\| = 1$ we have,

$$\left\|\sum_{i=1}^{m} F(f_i)g_i\right\| \leq \sum_{i=1}^{m}\|F(f_i)g_i\| = \sum_{i=1}^{m} |F(f_i)| \; \|g_i\| \leq \|F\| \sum_{i=1}^{m}\|f_i\| \, \|g_i\| .$$

By Lemma 1.2 the extreme left of the last inequality is invariant under equivalence. Thus, Definition 2.5 gives

$$\left\|\sum_{i=1}^{m} F(f_i)g_i\right\| \leq \gamma\left(\sum_{i=1}^{m} f_i \otimes g_i\right) \qquad \text{for all } F \in \mathcal{B}_1^* \text{ with } \|F\| = 1 .$$

Now suppose that $\sum_{i=1}^{m} f_i \otimes g_i$ is not equivalent to $0 \otimes 0$. By Lemma 1.7 there exists an $F \in \mathcal{B}_1^*$ for which $\sum_{i=1}^{m} F(f_i)g_i \neq 0$. Naturally we may suppose that $\|F\| = 1$. Consequently, $\gamma\left(\sum_{i=1}^{m} f_i \otimes g_i\right) > 0$.

II is immediate and IV is a consequence of Definition 2.5 for γ .

III: Let $\sum_{i=1}^{m} f_i \otimes g_i$ and $\sum_{i=1}^{n} h_i \otimes k_i$ be two given expressions and $\varepsilon > 0$. Clearly, we can find an expression

$$\sum_{i=1}^{r} f_i' \otimes g_i' \simeq \sum_{i=1}^{m} f_i \otimes g_i$$

such that

$$\sum_{i=1}^{r}\|f_i'\| \, \|g_i'\| < \gamma\left(\sum_{i=1}^{m} f_i \otimes g_i\right) + \varepsilon .$$

Similarly we can find

$$\sum_{i=1}^{s} h_i' \otimes k_i' \simeq \sum_{i=1}^{n} h_i \otimes k_i$$

such that

$$\sum_{i=1}^{s}\|h_i'\| \, \|k_i'\| < \gamma\left(\sum_{i=1}^{n} h_i \otimes k_i\right) + \varepsilon .$$

We have,

$$\sum_{i=1}^{r} f_i' \otimes g_i' \cdot + \cdot \sum_{i=1}^{s} h_i' \otimes k_i' \simeq \sum_{i=1}^{m} f_i \otimes g_i \cdot + \cdot \sum_{i=1}^{n} h_i \otimes k_i \ .$$

Condition IV and Definition 2.4 give,

$$\gamma\left(\sum_{i=1}^{m} f_i \otimes g_i \cdot + \cdot \sum_{i=1}^{n} h_i \otimes k_i\right) = \gamma\left(\sum_{i=1}^{r} f_i' \otimes g_i' \cdot + \cdot \sum_{i=1}^{s} h_i' \otimes k_i'\right) \leq$$

$$\sum_{i=1}^{r} \| f_i' \| \, \| g_i' \| + \sum_{i=1}^{s} \| h_i' \| \, \| k_i' \| \leq \gamma\left(\sum_{i=1}^{m} f_i \otimes g_i\right) + \gamma\left(\sum_{i=1}^{n} h_i \otimes k_i\right) + 2\varepsilon$$

The last inequality holds for any $\varepsilon > 0$. This proves III.

V: Let $\sum_{i=1}^{m} f_i \otimes g_i \simeq f \otimes g$. We choose an $F \in \mathfrak{B}_1^*$ such that $F(f) = \| f \|$ and $\| F \| = 1$ [1, p. 55]. By Lemma 1.2,

$$F(f) g = \sum_{i=1}^{m} F(f_i) g_i \ .$$

Consequently,

$$\| f \| \, \| g \| = \| F(f) g \| = \left\| \sum_{i=1}^{m} F(f_i) g_i \right\| \leq \sum_{i=1}^{m} \| f_i \| \, \| g_i \| \ .$$

Thus, Definition 2.5 furnishes $\gamma(f \otimes g) = \| f \| \, \| g \|$.

VI: Let S and T denote two operators on $\mathfrak{B}_1$ and $\mathfrak{B}_2$ respectively. Then,

$$\gamma\left(\sum_{i=1}^{m} Sf_i \otimes Tg_i\right) \leq \sum_{i=1}^{m} \| Sf_i \| \, \| Tg_i \| \leq$$

$$||| S ||| \; ||| T ||| \left(\sum_{i=1}^{m} \| f_i \| \, \| g_i \|\right) .$$

By Lemma 1.6, $\sum_{i=1}^{m} f_i \otimes g_i \simeq \sum_{j=1}^{m} h_j \otimes k_j$ implies $\sum_{i=1}^{m} Sf_i \otimes Tg_i \simeq \sum_{j=1}^{m} Sh_j \otimes Tk_j$ and therefore $\gamma\left(\sum_{i=1}^{m} Sf_i \otimes Tg_i\right) = \gamma\left(\sum_{j=1}^{m} Sh_j \otimes Tk_j\right)$. It follows that while we are running through the set of all expressions equivalent to $\sum_{i=1}^{m} f_i \otimes g_i$ the extreme left of the last inequality remains invariant. Consequently, Definition 2.5 gives

$$\gamma\left(\sum_{i=1}^{m} Sf_i \otimes Tg_i\right) \leq ||| S ||| \; ||| T ||| \; \gamma\left(\sum_{i=1}^{m} f_i \otimes g_i\right) .$$

This concludes the proof.

THEOREM 2.3. γ is the greatest crossnorm.

Proof. That γ is a crossnorm was proved in Lemma 2.13. Now, let α denote any crossnorm and $\sum_{i=1}^{m} f_i \otimes g_i$ a fixed expression. Then for any expression

$$\sum_{j=1}^{m} f_j' \otimes g_j' \simeq \sum_{i=1}^{m} f_i \otimes g_i$$

we have

$$\alpha(\sum_{i=1}^{m} f_i \otimes g_i) = \alpha(\sum_{j=1}^{m} f_j' \otimes g_j') \leqslant$$
$$\sum_{j=1}^{m} \alpha(f_j' \otimes g_j') = \sum_{j=1}^{m} \|f_j'\| \|g_j'\| .$$

Thus, Definition 2.5 gives,

$$\alpha(\sum_{i=1}^{m} f_i \otimes g_i) \leqslant \gamma(\sum_{i=1}^{m} f_i \otimes g_i) .$$

This concludes the proof.

REMARK 2.3. We notice that γ (as in the case of λ) is of "general character", that is, for any two given Banach spaces $\mathfrak{B}_1$ and $\mathfrak{B}_2$ without any special restrictions the unique greatest crossnorm can be always constructed In general, however, γ is not of "local character", that is, a lemma for γ analogous to that of Lemma 2.12 (proven for λ) is not true. Later however, we shall point out the precise conditions on the spaces $\mathfrak{B}_1$, $\mathfrak{B}_2$, for which the local character of γ is preserved.

For our further discussion we shall need the following simple propositio

LEMMA 2.14. Let $a_1, a_2, \ldots\ldots, a_n$ denote n real numbers and $b_1, b_2, \ldots\ldots, b_n$ denote n positive numbers. Then,

$$\frac{a_1 + a_2 + \ldots. + a_n}{b_1 + b_2 + \ldots. + b_n} \leqslant \max_{1 \leq i \leq n} \frac{a_i}{b_i}$$

Proof. The proof can be carried out easily by induction, verifying our Lemma first for n = 2 .

THEOREM 2.4. Let $\mathfrak{F}$ be a functional of expressions on $\mathfrak{B}_1 \odot \mathfrak{B}_2$ satisfying the following conditions:

(i). For equivalent expressions $\mathfrak{F}$ assumes the same value.

(ii). $\left|\mathfrak{F}(\sum_{i=1}^m f_i \otimes g_i)\right| \leq \sum_{i=1}^m \left|\mathfrak{F}(f_i \otimes g_i)\right|$.

Then,

$$\sup_{\sum_{i=1}^m f_i \otimes g_i} \frac{\left|\mathfrak{F}(\sum_{i=1}^m f_i \otimes g_i)\right|}{\gamma(\sum_{i=1}^m f_i \otimes g_i)} = \sup_{f,g} \frac{\left|\mathfrak{F}(f \otimes g)\right|}{\|f\| \, \|g\|} .$$

Proof. Since $\gamma(f \otimes g) = \|f\| \|g\|$ (Lemma 2.13, V) the right side of our equality is clearly not greater then the left side. Now let $\sum_{i=1}^m f_i \otimes g_i$ be fixed. Then for any expression $\sum_{j=1}^m f_j' \otimes g_j' \simeq \sum_{i=1}^m f_i \otimes g_i$, (i) and (ii) for $\mathfrak{F}$ furnish:

$$\frac{\left|\mathfrak{F}(\sum_{i=1}^m f_i \otimes g_i)\right|}{\sum_{j=1}^m \|f_j'\| \|g_j'\|} = \frac{\left|\mathfrak{F}(\sum_{j=1}^m f_j' \otimes g_j')\right|}{\sum_{j=1}^m \|f_j'\| \|g_j'\|} \leq \frac{\sum_{j=1}^m \left|\mathfrak{F}(f_j' \otimes g_j')\right|}{\sum_{j=1}^m \|f_j'\| \|g_j'\|}$$

By Lemma 2.14, the extreme right is

$$\leq \max_{1 \leq j \leq m} \frac{\left|\mathfrak{F}(f_j' \otimes g_j')\right|}{\|f_j'\| \|g_j'\|} \leq \sup_{f,g} \frac{\left|\mathfrak{F}(f \otimes g)\right|}{\|f\| \|g\|} .$$

This by Definition 2.5 implies

$$\frac{\left|\mathfrak{F}(\sum_{i=1}^m f_i \otimes g_i)\right|}{\gamma(\sum_{i=1}^m f_i \otimes g_i)} = \frac{\left|\mathfrak{F}(\sum_{i=1}^m f_i \otimes g_i)\right|}{\inf \sum_{j=1}^m \|f_j'\| \|g_j'\|} \leq \sup_{f,g} \frac{\left|\mathfrak{F}(f \otimes g)\right|}{\|f\| \|g\|} .$$

The last inequality holds for every expression $\sum_{i=1}^m f_i \otimes g_i$. Therefore, also the left side of the equality of our theorem is not greater than the right side. This concludes the proof.

THEOREM 2.5. The associate with the greatest crossnorm γ (Definition 2.5) on $\mathfrak{B}_1 \odot \mathfrak{B}_2$ is λ (Definition 2.4) on $\mathfrak{B}_1^* \odot \mathfrak{B}_2^*$, that is, $\gamma' = \lambda$.

Proof. For a fixed expression $\sum_{j=1}^{m} F_j \otimes G_j$ in $\mathcal{B}_1^* \odot \mathcal{B}_2^*$ the number

$$\left| \left(\sum_{j=1}^{m} F_j \otimes G_j\right)\left(\sum_{i=1}^{n} f_i \otimes g_i\right) \right|$$

obviously represents a functional of expressions $\sum_{i=1}^{n} f_i \otimes g_i$ in $\mathcal{B}_1 \odot \mathcal{B}_2$ satisfying the condition of Theorem 2.4. By Definition 2.2,

$$\gamma'\left(\sum_{j=1}^{m} F_j \otimes G_j\right) = \sup_{\sum_{i=1}^{n} f_i \otimes g_i} \frac{\left|\left(\sum_{j=1}^{m} F_j \otimes G_j\right)\left(\sum_{i=1}^{n} f_i \otimes g_i\right)\right|}{\gamma\left(\sum_{i=1}^{n} f_i \otimes g_i\right)}$$

By Theorem 2.4, the right side equals

$$\sup_{f,g} \frac{\left|\left(\sum_{j=1}^{m} F_j \otimes G_j\right)(f \otimes g)\right|}{\| f \| \, \| g \|}$$

The last number represents $\lambda\left(\sum_{j=1}^{m} F_j \otimes G_j\right)$ by Lemma 2.10. This concludes the proof.

CHAPTER III

CROSS-SPACES OF OPERATORS

1. The Banach spaces $\mathfrak{B}_1 \otimes_\alpha \mathfrak{B}_2$ and $\mathfrak{B}_1^* \otimes_{\alpha'} \mathfrak{B}_2^*$.

For a given norm (or crossnorm) α, the normed linear space $\mathfrak{B}_1 \odot_\alpha \mathfrak{B}_2$ in general, will not be complete. In that case we "complete" it that is, imbed it into the smallest possible Banach space in the usual Cantor-Meray fashion by adding new elements. We consider namely all fundamental sequences (that is, those which satisfy Cauchy's condition) of elements (or expressions representing those elements) in $\mathfrak{B}_1 \odot_\alpha \mathfrak{B}_2$ and introduce the following standard identifications [5, p. 106]:

(i) A sequence consisting of identical elements we identify with that element.

(ii) Two fundamental sequences of elements we consider identical if and only if, the norm of their difference tends toward 0.

(iii) The norm of a fundamental sequence of elements is defined as the limit of the norms of the elements in that sequence.

DEFINITION 3.1. The Cantor-Meray completion of the normed linear space $\mathfrak{B}_1 \odot_\alpha \mathfrak{B}_2$ which naturally depends on the norm α, will be denoted by $\mathfrak{B}_1 \otimes_\alpha \mathfrak{B}_2$ and termed a "direct product" of $\mathfrak{B}_1$ and $\mathfrak{B}_2$. Whenever α is a crossnorm, $\mathfrak{B}_1 \otimes_\alpha \mathfrak{B}_2$ is termed a "cross-space". Similarly, completing $\mathfrak{B}_1^* \odot_{\alpha'} \mathfrak{B}_2^*$ we obtain $\mathfrak{B}_1^* \otimes_{\alpha'} \mathfrak{B}_2^*$ and term it the space associated with or the associate space for the cross-space $\mathfrak{B}_1 \otimes_\alpha \mathfrak{B}_2$.

REMARK 3.1. The associate space is defined whenever α' is always finite. This is always the case whenever, $\alpha \geqq \lambda$.

2. The inclusion $\mathfrak{B}_1^* \otimes_{\alpha'} \mathfrak{B}_2^* \subset (\mathfrak{B}_1 \otimes_\alpha \mathfrak{B}_2)^*$.

Let α denote a crossnorm $\geqq \lambda$ on $\mathfrak{B}_1 \odot \mathfrak{B}_2$ (Definition 2.4). By Theorem 2.1, α' is also a crossnorm on $\mathfrak{B}_1^* \odot \mathfrak{B}_2^*$. Thus, two given Banach spaces $\mathfrak{B}_1$, $\mathfrak{B}_2$, and a given crossnorm $\alpha \geqq \lambda$ determine the spaces $\mathfrak{B}_1 \otimes_\alpha \mathfrak{B}_2$ and $\mathfrak{B}_1^* \otimes_{\alpha'} \mathfrak{B}_2^*$.

It is a consequence of Definition 2.2 and Theorem 2.1 that for a fixed expression $\sum_{j=1}^m F_j \otimes G_j$ the "inner product" $(\sum_{j=1}^m F_j \otimes G_j)(\sum_{i=1}^n f_i \otimes g_i)$ represents an additive functional on $\mathfrak{B}_1 \odot_\alpha \mathfrak{B}_2$ with a bound equal to $\alpha'(\sum_{j=1}^m F_j \otimes G_j)$. Since by definition $\mathfrak{B}_1 \odot_\alpha \mathfrak{B}_2$ is dense in $\mathfrak{B}_1 \otimes_\alpha \mathfrak{B}_2$, this functional can be extended in a unique manner to an additive functional on $\mathfrak{B}_1 \otimes_\alpha \mathfrak{B}_2$, without affecting the value of its bound. We may write therefore,

$$(\mathfrak{B}_1 \otimes_\alpha \mathfrak{B}_2)^* \supset \mathfrak{B}_1^* \otimes_{\alpha'} \mathfrak{B}_2^* .$$

REMARK 3.2. The last inclusion represents more than merely the equivalence of $\mathfrak{B}_1^* \otimes_{\alpha'} \mathfrak{B}_2^*$ with a certain subspace of $(\mathfrak{B}_1 \otimes_\alpha \mathfrak{B}_2)^*$ as defined in [1, p. 180] , since that equivalence is to be understood in the sense explained above. Throughout the rest of this paper all inclusions of this form will have to be understood in the sense explained above.

It does not appear to be a simple task to state the precise conditions imposed upon a crossnorm α for which the resulting cross-space $\mathfrak{B}_1 \otimes_\alpha \mathfrak{B}_2$

is such that its conjugate space coincides with its associate space. This is always the case, for instance, when at least one of the spaces $\mathfrak{B}_1$ or $\mathfrak{B}_2$ is finite dimensional. We present the first step in this direction by interpreting $(\mathfrak{B}_1 \otimes_\alpha \mathfrak{B}_2)^*$ as well as $\mathfrak{B}_1^* \otimes_{\alpha'} \mathfrak{B}_2^*$ as Banach spaces of some (not necessarily all) operators from $\mathfrak{B}_1$ into $\mathfrak{B}_2^*$ (from $\mathfrak{B}_2$ into $\mathfrak{B}_1^*$), where the norm of an operator is in general different from its bound. Furthermore, we are able to present a complete discussion for the case when we deal with the greatest crossnorm γ (Definition 2.5 and Theorem 2.3).

3. $(\mathfrak{B}_1 \otimes_\alpha \mathfrak{B}_2)^*$ as the space of operators of finite α-norm.

Throughout the rest of this Chapter let α stand for a crossnorm on $\mathfrak{B}_1 \odot \mathfrak{B}_2$ although this is not stated each time explicitly.

DEFINITION 3.2. An operator A from $\mathfrak{B}_1$ into $\mathfrak{B}_2^*$ is termed of finite α-norm, if there exists a finite constant C such that

$$\left| \sum_{i=1}^{n} (Af_i) g_i \right| \leq C\,\alpha\left(\sum_{i=1}^{n} f_i \otimes g_i\right)$$

for all expressions $\sum_{i=1}^{n} f_i \otimes g_i$ in $\mathfrak{B}_1 \odot \mathfrak{B}_2$. The least of such constants we denote by $\|A\|_\alpha$ and term the " α-norm" of A. For an operator A for which such a finite constant does not exist, we define $\|A\|_\alpha = +\infty$.

The justification of such a definition will be clear from Theorem 3.1 which follows.

LEMMA 3.1. The operators A of finite α-norm form a normed linear space if $\|A\|_\alpha$ represents the norm of A.

Proof. The proof does not present any difficulty.

LEMMA 3.2. For an operator A from $\mathfrak{B}_1$ into $\mathfrak{B}_2^*$ we have always

$$\| A \|_\alpha \geq |||A|||$$

for any crossnorm α.

Proof. For every pair f, g, Definition 3.2 gives

$$|(Af)g| \leq \| A \|_\alpha \, \alpha(f \otimes g) = \| A \|_\alpha \| f \| \| g \| .$$

Thus, $|||A||| \leq \| A \|_\alpha$. This concludes the proof.

THEOREM 3.1. $(\mathfrak{B}_1 \otimes_\alpha \mathfrak{B}_2)^*$ may be interpreted as the Banach space of all operators A from $\mathfrak{B}_1$ into $\mathfrak{B}_2^*$ of finite α-norm, where $\| A \|_\alpha$ represents the norm of A. In other words:

An element $\mathfrak{F}$ of $(\mathfrak{B}_1 \otimes_\alpha \mathfrak{B}_2)^*$ generates an operator A from $\mathfrak{B}_1$ into $\mathfrak{B}_2^*$ of finite α-norm and conversely. This correspondence is such that:

(1) $\mathfrak{F} \rightleftarrows A$ implies $a\mathfrak{F} \rightleftarrows aA$.

(2) $\mathfrak{F}_1 \rightleftarrows A_1$ and $\mathfrak{F}_2 \rightleftarrows A_2$ implies $\mathfrak{F}_1 + \mathfrak{F}_2 \rightleftarrows A_1 + A_2$.

(3) $|||\mathfrak{F}||| = \| A \|_\alpha$ whenever $\mathfrak{F} \rightleftarrows A$.

Proof. Let $\mathfrak{F} \in (\mathfrak{B}_1 \otimes_\alpha \mathfrak{B}_2)^*$ and $|||\mathfrak{F}|||$ denote its bound. Clearly,

(i) $\mathfrak{F}(a.f \otimes g) = \mathfrak{F}((af) \otimes g) = \mathfrak{F}(f \otimes (ag)) = a.\mathfrak{F}(f \otimes g)$

for any scalar a.

(ii) $\mathfrak{F}(f_1 \otimes g_1 + f_2 \otimes g_2) = \mathfrak{F}(f_1 \otimes g_1) + \mathfrak{F}(f_2 \otimes g_2)$.

In particular,

(ii') $\mathfrak{F}((f_1 + f_2) \otimes g) = \mathfrak{F}(f_1 \otimes g) + \mathfrak{F}(f_2 \otimes g)$.

(ii'') $\mathfrak{F}(f \otimes (g_1 + g_2)) = \mathfrak{F}(f \otimes g_1) + \mathfrak{F}(f \otimes g_2)$.

Relations (i) and (ii'') prove that for a fixed f_0, $\mathcal{F}(f_0 \otimes g)$ represents an additive functional of g. That functional is bounded since

$$|\mathcal{F}(f_0 \otimes g)| \leq |||\mathcal{F}|||\, \alpha(f_0 \otimes g) = (|||\mathcal{F}|||\, \|f_0\|)\, \|g\| \qquad \text{for } g \in \mathcal{B}_2.$$

Therefore, $\mathcal{F}$ assigns to every $f \in \mathcal{B}_1$ a unique element of $\mathcal{B}_2^*$ which we shall denote by Af, so that

$$\mathcal{F}(f \otimes g) = (Af)g \qquad \text{for } f \in \mathcal{B}_1, \quad g \in \mathcal{B}_2$$

and therefore by (ii)

$$\text{(a)} \qquad \mathcal{F}\left(\sum_{i=1}^{m} f_i \otimes g_i\right) = \sum_{i=1}^{m} \mathcal{F}(f_i \otimes g_i) = \sum_{i=1}^{m} (Af_i)g_i \ .$$

Thus, A is additive.

Since $\mathcal{B}_1 \odot_\alpha \mathcal{B}_2$ is dense in $\mathcal{B}_1 \otimes_\alpha \mathcal{B}_2$, the bound of $\mathcal{F}$ on both spaces is the same, and thus $|||\mathcal{F}|||$ represents the least constant C satisfying the inequality:

$$\text{(b)} \qquad \left|\sum_{i=1}^{m} (Af_i)g_i\right| = \left|\mathcal{F}\left(\sum_{i=1}^{m} f_i \otimes g_i\right)\right| \leq C\, \alpha\left(\sum_{i=1}^{m} f_i \otimes g_i\right)$$

for all expressions $\sum_{i=1}^{m} f_i \otimes g_i$ in $\mathcal{B}_1 \odot \mathcal{B}_2$.

By Definition 3.2 however, that least constant C satisfying (b) was denoted by $\|A\|_\alpha$. Therefore, $\|A\|_\alpha = |||\mathcal{F}||| < +\infty$. A is bounded since $|||A||| \leq \|A\|_\alpha$.

Thus, an $\mathcal{F}$ in $(\mathcal{B}_1 \otimes_\alpha \mathcal{B}_2)^*$ determines a unique operator A from $\mathcal{B}_1$ into $\mathcal{B}_2^*$ of finite α-norm for which $\|A\|_\alpha = |||\mathcal{F}|||$.

Conversely. For an operator A from $\mathcal{B}_1$ into $\mathcal{B}_2^*$ with $\|A\|_\alpha < +\infty$ we construct $\mathcal{F}$ by means of (a) above. $\mathcal{F}$ is uniquely defined on $\mathcal{B}_1 \odot_\alpha \mathcal{B}_2$ since by Lemma 1.5, $\sum_{i=1}^{m} (Af_i)g_i$ is invariant under equivalence. Then, again by (b) above, $\mathcal{F}$ is bounded on $\mathcal{B}_1 \odot_\alpha \mathcal{B}_2$ hence on $\mathcal{B}_1 \otimes_\alpha \mathcal{B}_2$ and

$|||\mathcal{G}||| = \|A\|_\alpha$. The correspondence $\mathcal{G} \rightleftarrows A$ is obviously additive. This concludes the proof.

COROLLARY 3.1. The space of all operators A of finite α-norm is complete in the norm $\|A\|_\alpha$.

Proof. The proof is a consequence of Theorem 3.1.

4. The space of all operators.

LEMMA 3.3. Every operator A from $\mathcal{B}_1$ into $\mathcal{B}_2^*$ is of finite γ-norm. Moreover, $\|A\|_\gamma = |||A|||$.

Proof. For an operator A , the sum $\sum_{i=1}^m (Af_i)g_i$ is invariant under equivalence (Lemma 1.5). Consequently, $|\sum_{i=1}^m (Af_i)g_i|$ is a functional of expressions satisfying the assumptions of Theorem 2.4. By that theorem,

$$\|A\|_\gamma = \sup_{\sum_{i=1}^m f_i \otimes g_i} \frac{|\sum_{i=1}^m (Af_i)g_i|}{\gamma(\sum_{i=1}^m f_i \otimes g_i)} = \sup_{f,g} \frac{|(Af)g|}{\|f\|\,\|g\|}$$

The extreme right clearly represents the bound of A . This concludes the proof.

THEOREM 3.2. $(\mathcal{B}_1 \otimes_\gamma \mathcal{B}_2)^*$ may be interpreted as the Banach space of all operators from $\mathcal{B}_1$ into $\mathcal{B}_2^*$, where the norm of an operator equals to its bound.

Proof. The proof is a consequence of Theorem 3.1 and Lemma 3.3.

5. A "natural equivalence".

REMARK 3.3. Let $\mathfrak{F} \in (\mathcal{B}_1 \otimes_\alpha \mathcal{B}_2)^*$. By Defition 3.2 and Theorem 3.1 $\mathfrak{F}(f \otimes g)$ determines an operator A from $\mathcal{B}_1$ into $\mathcal{B}_2^*$ of finite α-norm for which $\| A \|_\alpha = |||\mathfrak{F}|||$. The same $\mathfrak{F}$ however, determines a "corresponding" operator $\tilde{A}$ from $\mathcal{B}_2$ into $\mathcal{B}_1^*$ for which $\| \tilde{A} \|_\alpha = |||\mathfrak{F}|||$. This can be readily seen as follows: For a fixed $g_0 \in \mathcal{B}_2$, $\mathfrak{F}(f \otimes g_0)$ represents an additive and bounded functional on $\mathcal{B}_1$. Thus, $\mathfrak{F}(f \otimes g)$ also determines a unique $\tilde{A}$ from $\mathcal{B}_2$ into $\mathcal{B}_1^*$. Furthermore,

$$\mathfrak{F}(f \otimes g) = (Af)g = (\tilde{A}g)f \qquad \text{for } f \in \mathcal{B}_1,\ g \in \mathcal{B}_2.$$

That $\| \tilde{A} \|_\alpha = |||\mathfrak{F}|||$, is shown in a manner analogous to $\| A \|_\alpha = |||\mathfrak{F}|||$ in the proof of Theorem 3.1.

For "corresponding" operators of finite α-norm (that is, generated by the same $\mathfrak{F}$) in addition to $\| A \|_\alpha = \| \tilde{A} \|_\alpha$ we also have $||| A ||| = ||| \tilde{A} |||$. This can be seen as follows:

$$||| A ||| = \sup_{\|f\|=1} \| Af \| = \sup_{\|f\|=\|g\|=1} |\mathfrak{F}(f \otimes g)| =$$
$$\sup_{\|g\|=1} \| \tilde{A}g \| = ||| \tilde{A} ||| .$$

Thus, all preceding Lemmas and Theorems of this section are also valid if in their wording we replace the phrase "operator from $\mathcal{B}_1$ into $\mathcal{B}_2^*$" by "operator from $\mathcal{B}_2$ into $\mathcal{B}_1^*$". In particular $(\mathcal{B}_1 \otimes_\alpha \mathcal{B}_2)^*$ may be regarded as the Banach space of all operators from $\mathcal{B}_2$ into $\mathcal{B}_1^*$ of finite α-norm. In general the corresponding operators A, $\tilde{A}$, are such that one is a contraction of the adjoint of the other. For reflexive Banach spaces the corresponding operators are adjoint to each other.

This is expressed in detail in the following theorem:

THEOREM 3.3. There exists a "natural equivalence $\rightleftarrows$ " between the Banach space $\mathfrak{B}$ of all operators A from $\mathfrak{B}_1$ into $\mathfrak{B}_2^*$ of finite α-norm and the Banach space $\widetilde{\mathfrak{B}}$ of all operators $\widetilde{A}$ from $\mathfrak{B}_2$ into $\mathfrak{B}_1^*$ of finite α-norm in the sense that,

(i) $a_1A_1 + a_2A_2 \rightleftarrows a_1\widetilde{A}_1 + a_2\widetilde{A}_2$ (a_1, a_2, are real numbers).

(ii) $A^* \supset \widetilde{A}$ and $\widetilde{A}^* \supset A$.

(iii) $\| A \|_\alpha = \| \widetilde{A} \|_\alpha$.

(iv) $||| A ||| = ||| \widetilde{A} |||$.

Proof. The proof is a consequence of Remark 3.1.

6. $\mathfrak{B}_1^* \otimes_{\alpha'} \mathfrak{B}_2^*$ as a space of operators.

LEMMA 3.4. For a given crossnorm $\alpha \geqslant \lambda$, the operators of finite rank are of finite α-norm. The normed linear space of operators A from $\mathfrak{B}_1$ into $\mathfrak{B}_2^*$ of finite rank, where $\| A \|_\alpha$ represents the norm, may be characterized as $\mathfrak{B}_1^* \odot_{\alpha'} \mathfrak{B}_2^*$.

Proof. A given expression $\Sigma_{j=1}^m F_j \otimes G_j$ corresponds in the manner indicated in the proof of Theorem 3.1 to an operator A from $\mathfrak{B}_1$ into $\mathfrak{B}_2^*$ defined by the relation $Af = \Sigma_{j=1}^m F_j(f)G_j$. Conversely, every operator from $\mathfrak{B}_1$ into $\mathfrak{B}_2^*$ of finite rank can be obtained in such a manner. The correspondence is clearly additive. Since $\alpha \geqslant \lambda$, the value $\alpha'(\Sigma_{j=1}^m F_j \otimes G_j)$ is finite by Theorem 2.1. Since,

$$(\Sigma_{j=1}^m F_j \otimes G_j)(\Sigma_{i=1}^n f_i \otimes g_i) = \Sigma_{i=1}^n (Af_i)g_i \quad ,$$

Definitions 2.2 and 3.2 furnish

$$\alpha'(\Sigma_{j=1}^{m} F_j \otimes G_j) = \|A\|_\alpha$$

This concludes the proof.

THEOREM 3.4. Let α be a given crossnorm $\geq \lambda$. Then, $\mathcal{B}_1^* \otimes_{\alpha'} \mathcal{B}_2^*$ may be interpreted as the Banach space of all operators A from $\mathcal{B}_1$ into $\mathcal{B}_2^*$ of finite α-norm (with $\|A\|_\alpha$ representing the norm of A), which may be approximated in that norm by operators of finite rank.

Proof. A given element in $\mathcal{B}_1^* \otimes_{\alpha'} \mathcal{B}_2^*$ is represented by a sequence of expressions

$$\Sigma_{j=1}^{m_p} F_j^{(p)} \otimes G_j^{(p)}$$

in $\mathcal{B}_1^* \odot_{\alpha'} \mathcal{B}_2^*$ fundamental relative to α'; the norm of such an element equals to the limit of the norms of the expressions in the fundamental sequence representing it. By Lemma 3.4 such a sequence of expressions determines a sequence A_p of operators of finite rank from $\mathcal{B}_1$ into $\mathcal{B}_2^*$ for which

$$\lim_{p,q\to\infty} \|A_p - A_q\|_\alpha = 0 .$$

By Corollary 3.1, the space of operators of finite α-norm is complete. Hence A_p determines an operator A of finite α-norm for which

$$\lim_{p\to\infty} \|A - A_p\|_\alpha = 0 .$$

By Lemma 3.2, A is uniquely determined; it may be identified with the given element. Furthermore,

$$\lim_{p\to\infty} \alpha'(\Sigma_{j=1}^{m_p} F_j^{(p)} \otimes G_j^{(p)}) = \lim_{p\to\infty} \|A_p\|_\alpha = \|A\|_\alpha .$$

This concludes the proof.

THEOREM 3.5. $\mathcal{B}_1^* \otimes_{\gamma'} \mathcal{B}_2^* = \mathcal{B}_1^* \otimes_\lambda \mathcal{B}_2^*$ represents the Banach space of operators from $\mathcal{B}_1$ into $\mathcal{B}_2^*$ approximable in bound by operators of finite rank.

Proof. That $\gamma' = \lambda$ was proven in Theorem 2.5. The rest is a consequence of Lemma 3.3 and Theorem 3.4.

THEOREM 3.6. Let α denote a crossnorm $\geq \lambda$. The equality $(\mathfrak{B}_1 \otimes_\alpha \mathfrak{B}_2)^* = \mathfrak{B}_1^* \otimes_{\alpha'} \mathfrak{B}_2^*$ occurs if and only if, every operator from $\mathfrak{B}_1$ into $\mathfrak{B}_2^*$ of finite α-norm can be approximated in that norm by operators from $\mathfrak{B}_1$ into $\mathfrak{B}_2^*$ of finite rank.

Proof. The proof is a consequence of Theorem 3.1 and Theorem 3.4.

LEMMA 3.5. For any crossnorm $\alpha \geq \lambda$, the operators determined by $\mathfrak{B}_1^* \otimes_{\alpha'} \mathfrak{B}_2^*$ are completely continuous [1, p. 96] .

Proof. Let A be an operator from $\mathfrak{B}_1$ into $\mathfrak{B}_2^*$ determined by $\mathfrak{B}_1^* \otimes_{\alpha'} \mathfrak{B}_2^*$. Thus, for some sequence of operators A_p of finite rank we have,

$$\lim_{p \to \infty} \| A - A_p \|_\alpha = 0 .$$

By Lemma 3.2,

$$\lim_{p \to \infty} ||| A - A_p ||| \leq \lim \| A - A_p \|_\alpha = 0 .$$

Thus, A is completely continuous by [1, p. 96, Theorem 2].

REMARK 3.4. An example for which $\mathfrak{B}_1^* \otimes_\lambda \mathfrak{B}_2^*$ is a proper subspace of the Banach space of all completely continuous operators from $\mathfrak{B}_1$ into $\mathfrak{B}_2^*$ would answer negatively the "basis problem" [1, p. 111] . Suppose A is a completely continuous operator which cannot be approximated in bound by operators of finite rank. The range of A may be considered contained in a separable Banach space $\mathfrak{B}$ [1, p. 96, Theorem 1] . It is well known that a completely continuous operator on a Banach space whose values lie in a

Banach space having a basis, can be always approximated in bound by operators of finite rank [11]. It follows that $\mathfrak{B}$ must not have a basis. Otherwise, A is approximable in bound by operators of finite rank.

THEOREM 3.7. Let α denote a crossnorm $\gg \lambda$. If there exists an operator of finite α-norm which is not completely continuous, then $\mathfrak{B}_1^* \otimes_{\alpha'} \mathfrak{B}_2^*$ is a proper subspace of $(\mathfrak{B}_1 \otimes_\alpha \mathfrak{B}_2)^*$.

Proof. The proof is a consequence of Lemma 3.5 and Theorem 3.1.

At this point we are ready to supplement Theorem 3.3 with the following assertion:

THEOREM 3.8. Let α be a crossnorm $\gg \lambda$. The "natural equivalence" of Theorem 3.3 is such that it also preserves a natural equivalence between the Banach space of all operators from $\mathfrak{B}_1$ into $\mathfrak{B}_2^*$ of finite α-norm which are approximable in that norm by operators of finite rank, and the Banach space of all operators from $\mathfrak{B}_2$ into $\mathfrak{B}_1^*$ of finite α-norm which may be approximated in that norm by operators of finite rank.

Proof. An expression $\sum_{j=1}^{m} F_j \otimes G_j$ in $\mathfrak{B}_1^* \odot \mathfrak{B}_2^*$ determines "corresponding" operators $\sum_{j=1}^{m} F_j(f)G_j$, $\sum_{j=1}^{m} G_j(g)F_j$. The α-norm of both of these operators equals to $\alpha'(\sum_{j=1}^{m} F_j \otimes G_j)$. The rest follows from the arguments in the proof of Theorem 3.4.

7. The local character of γ as a characteristic property of unitary spaces.

We conclude this chapter by settling the "extension problem" for the greatest crossnorm γ , indicating at the same time an application of Theorem 3.2.

An operator P on a perfectly general Banach space is termed a projection if and only if, $P^2 = P$. The range of P is the closed linear manifold of all the f's for which $Pf = f$. The bound of such a projection may be > 1 . Closely connected with the notion of a projection is that of a complementary manifold. Let $\mathfrak{M}$ denote a closed linear manifold in a Banach space $\mathfrak{B}$. A closed linear manifold $\mathfrak{N}$ in $\mathfrak{B}$ is said to be complementary to $\mathfrak{M}$ if and only if, every f in $\mathfrak{B}$ can be uniquely represented in the form $f = f' + f''$, where $f' \in \mathfrak{M}$ and $f'' \in \mathfrak{N}$. It is readily seen, as is also shown for instance, in [7, p. 138, Lemma 1.1.1] that the existence of a projection P of $\mathfrak{B}$ on $\mathfrak{M}$ is equivalent to the existence in of a complementary manifold $\mathfrak{N}$ to $\mathfrak{M}$ (namely, $\mathfrak{N}$ is the set of all f's for which $Pf = 0$).

Let $\mathfrak{M}_1$ and $\mathfrak{M}_2$ denote two closed linear manifolds in $\mathfrak{B}_1$ and $\mathfrak{B}_2$ respectively. Let $\sum_{i=1}^{n} f_i \otimes g_i$ be a fixed expression in $\mathfrak{M}_1 \odot \mathfrak{M}_2 \subset \mathfrak{B}_1 \odot \mathfrak{B}_2$. Obviously $\mathfrak{B}_1 \odot \mathfrak{B}_2$ possesses more expressions equivalent to $\sum_{i=1}^{n} f_i \otimes g_i$ than $\mathfrak{M}_1 \odot \mathfrak{M}_2$ does. By Definition 2.5 therefore, the value for $\sum_{i=1}^{n} f_i \otimes g_i$ of the greatest crossnorm on $\mathfrak{M}_1 \odot \mathfrak{M}_2$ is not smaller than the value for that expression of the greatest crossnorm on $\mathfrak{B}_1 \odot \mathfrak{B}_2$.

LEMMA 3.6. Let $\mathfrak{B}$ denote a Banach space and $\mathfrak{M}$ a closed linear manifold in $\mathfrak{B}$ such that $\mathfrak{M}$ is the conjugate space of some Banach space (that is, $\mathfrak{M} = \mathfrak{M}_0^*$). Suppose further that,

$$\mathfrak{M} \otimes_\gamma \mathfrak{M}_0 \subset \mathfrak{B} \otimes_\gamma \mathfrak{M}_0 ,$$

that is, the greatest crossnorm on $\mathfrak{B} \otimes_\gamma \mathfrak{M}_0$ is an "extension" of the greatest crossnorm on $\mathfrak{M} \otimes_\gamma \mathfrak{M}_0$. Then, there exists a projection of $\mathfrak{B}$ on $\mathfrak{M}$ with bound 1.

Proof. By Theorem 3.2, $(\mathfrak{M} \otimes_\gamma \mathfrak{M}_0)^*$ represents the Banach space of all operators from $\mathfrak{M}$ into $\mathfrak{M}_0^* = \mathfrak{M}$, where the norm of an operator is given by its bound on $\mathfrak{M}$. Similarly, $(\mathfrak{B} \otimes_\gamma \mathfrak{M}_0)^*$ represents the Banach space of all operators from $\mathfrak{B}$ into $\mathfrak{M}_0^* = \mathfrak{M}$. By Theorems 3.1 and 3.2, the identity operator P_0 from $\mathfrak{M}$ into $\mathfrak{M}$ generates an additive bounded functional $\mathfrak{F}_0$ on $\mathfrak{M} \otimes_\gamma \mathfrak{M}_0$ such that,

(1) $\mathfrak{F}_0(f \otimes g) = (P_0 f)g$ for $f \in \mathfrak{M}$, $g \in \mathfrak{M}_0$, and

(2) $|||\mathfrak{F}_0||| = |||P_0||| = 1$.

By assumption the space $\mathfrak{M} \otimes_\gamma \mathfrak{M}_0$ is a subspace of $\mathfrak{B} \otimes_\gamma \mathfrak{M}_0$. Thus, by Hahn-Banach's extension theorem [1, p. 55, Theorem 2], $\mathfrak{F}_0$ can be extended to an additive and bounded functional $\mathfrak{F}$ on $\mathfrak{B} \otimes_\gamma \mathfrak{M}_0$ without changing its bound, that is, for which

(3) $\mathfrak{F}(f \otimes g) = \mathfrak{F}_0(f \otimes g)$ for $f \in \mathfrak{M}$, $g \in \mathfrak{M}_0$, and

(4) $|||\mathfrak{F}||| = |||\mathfrak{F}_0|||$.

Again this $\mathfrak{F}$ generates an operator P from $\mathfrak{B}$ into $\mathfrak{M}_0^* = \mathfrak{M}$, for which

(5) $\mathfrak{F}(f \otimes g) = (Pf)g$ for $f \in \mathfrak{B}$, $g \in \mathfrak{M}_0$, and

(6) $|||\mathfrak{F}||| = |||P|||$.

Clearly, (1), (3) and (5) prove that P is an extension of $P_\circ$; the domain of P is $\mathfrak{B}$. Furthermore,

$$|||P||| = |||P_\circ||| = 1 ,$$

by virtue of (2), (4) and (6). Thus the bound of P equals 1 . Furthermore, since P is identical on $\mathfrak{M}$, the relation $P^2 = P$ holds, that is, P is a projection of $\mathfrak{B}$ on $\mathfrak{M}$. This concludes the proof.

LEMMA 3.7. Let $\mathfrak{M}$ denote a closed linear manifold in a Banach space $\mathfrak{B}$. Then, the existence of a projection of $\mathfrak{B}$ on $\mathfrak{M}$ with bound 1 implies that

$$\mathfrak{M} \otimes_\gamma \mathfrak{A} \subset \mathfrak{B} \otimes_\gamma \mathfrak{A}$$

for any Banach space $\mathfrak{A}$.

Proof. It is sufficient to prove that $\mathfrak{M} \odot_\gamma \mathfrak{A} \subset \mathfrak{B} \odot_\gamma \mathfrak{A}$. Let P be a projection of $\mathfrak{B}$ on $\mathfrak{M}$ with bound 1. We denote by $\mathfrak{N}$ the closed linear manifold of all elements f in $\mathfrak{B}$ for which $Pf = 0$. $\mathfrak{M}$ and $\mathfrak{N}$ are complementary. Let $\sum_{i=1}^{n} f_i^\circ \otimes g_i^\circ$ be a fixed expression in $\mathfrak{M} \odot \mathfrak{A}$, and $\sum_{j=1}^{m} f_j \otimes g_j$ any expression in $\mathfrak{B} \odot \mathfrak{A}$ equivalent to it. Then, $\sum_{j=1}^{m} Pf_j \otimes g_j$ which is an expression in $\mathfrak{M} \odot \mathfrak{A}$ is also equivalent to $\sum_{i=1}^{n} f_i^\circ \otimes g_i^\circ$. To prove this, put $f_j = f_j' + f_j''$ where $f_j' = Pf_j \in \mathfrak{M}$, $f_j'' \in \mathfrak{N}$ for $j = 1, 2, \ldots\ldots, m$. Thus,

$$\sum_{j=1}^{m} f_j \otimes g_j \simeq \sum_{j=1}^{m} Pf_j \otimes g_j \cdot + \cdot \sum_{j=1}^{m} f_j'' \otimes g_j ,$$

and therefore,

$$\sum_{i=1}^{n} f_i^\circ \otimes g_i^\circ \cdot + \cdot \sum_{j=1}^{m} (-Pf_j) \otimes g_j \simeq \sum_{j=1}^{m} f_j'' \otimes g_j .$$

The left side clearly, represents an expression in $\mathfrak{M} \odot \mathfrak{A}$, while the right side is an expression in $\mathfrak{N} \odot \mathfrak{A}$. Thus, both sides must be equivalent to $0 \otimes 0$

Therefore, $\sum_{i=1}^{m} f_i^\circ \otimes g_i^\circ \simeq \sum_{j=1}^{m} Pf_j \otimes g_j$.

Since P denotes a projection of $\mathfrak{B}$ on $\mathfrak{M}$ with bound 1, we have,

$$\sum_{j=1}^{m} \|f_j\| \, \|g_j\| \geqslant \sum_{j=1}^{m} \|Pf_j\| \, \|g_j\| \quad .$$

Thus, the inf in Definition 2.5 obtained by taking all expressions from $\mathfrak{B} \odot \mathfrak{A}$ equivalent to $\sum_{i=1}^{m} f_i^\circ \otimes g_i^\circ$, is not smaller than the one obtained by taking all expressions from $\mathfrak{M} \odot \mathfrak{A}$ equivalent to $\sum_{i=1}^{m} f_i^\circ \otimes g_i^\circ$. On the other hand the converse always holds. This was pointed out in an argument preceding Lemma 3.6. This concludes the proof.

COROLLARY 3.2. Let $\mathfrak{M}$ denote a closed linear manifold in a Banach space $\mathfrak{B}$. Suppose further that $\mathfrak{M}$ is the conjugate space of another Banach space $\mathfrak{M}_0$. Then, the inclusion $\mathfrak{M} \otimes_\gamma \mathfrak{M}_0 \subset \mathfrak{B} \otimes_\gamma \mathfrak{M}_0$ implies $\mathfrak{M} \otimes_\gamma \mathfrak{A} \subset \mathfrak{B} \otimes_\gamma \mathfrak{A}$ for any Banach space $\mathfrak{A}$.

Proof. The proof is a consequence of Lemmas 3.6 and 3.7.

COROLLARY 3.3. Let $\mathfrak{M}$ denote a closed linear manifold in a Hilbert space $\mathfrak{H}$. Then, $\mathfrak{M} \otimes_\gamma \mathfrak{A} \subset \mathfrak{H} \otimes_\gamma \mathfrak{A}$ for any Banach space $\mathfrak{A}$.

Proof. For a closed linear manifold $\mathfrak{M}$ in $\mathfrak{H}$ there always exists a projection of $\mathfrak{H}$ on $\mathfrak{M}$ with bound 1. An application of Corollary 3.2 concludes the proof.

COROLLARY 3.4. Let $\mathfrak{M}_1$ and $\mathfrak{M}_2$ denote two closed linear manifolds in the Banach spaces $\mathfrak{B}_1$ and $\mathfrak{B}_2$ respectively. Suppose further that there exists a projection of $\mathfrak{B}_1$ on $\mathfrak{M}_1$ with bound 1 and a projection of $\mathfrak{B}_2$ on $\mathfrak{M}_2$ with bound 1. Then,

$$\mathfrak{M}_1 \otimes_\gamma \mathfrak{M}_2 \subset \mathfrak{B}_1 \otimes_\gamma \mathfrak{B}_2 .$$

Proof. Applying twice Corollary 3.2, we get $\mathfrak{M}_1 \otimes_\gamma \mathfrak{M}_2 \subset \mathfrak{B}_1 \otimes_\gamma \mathfrak{M}_2$ and $\mathfrak{B}_1 \otimes_\gamma \mathfrak{M}_2 \subset \mathfrak{B}_1 \otimes_\gamma \mathfrak{B}_2$. This concludes the proof.

THEOREM 3.9. A Banach space $\mathfrak{B}$ is unitary if and only if,

$$\mathfrak{M} \otimes_\gamma \mathfrak{M}^* \subset \mathfrak{B} \otimes_\gamma \mathfrak{M}^*$$

for any two-dimensional linear manifold $\mathfrak{M} \subset \mathfrak{B}$.

Proof. We recall that a Banach space $\mathfrak{B}$ is unitary if and only if, its norm $\| f \|$ satisfies the relationship

$$\| f_1 + f_2 \|^2 + \| f_1 - f_2 \|^2 = 2 (\| f_1 \|^2 + \| f_2 \|^2)$$

for every pair f_1 and f_2 in $\mathfrak{B}$.

The proof is thus a consequence of Kakutani's characterization of unitary spaces which states that a Banach space is unitary if and only if, there exist projection operators of bound 1 on every two-dimensional linear manifold $\mathfrak{M} \subset \mathfrak{B}$, and of Lemmas 3.6 and 3.7.

COROLLARY 3.5. Whenever a Banach space $\mathfrak{B}$ is not unitary there exists a two-dimensional linear manifold $\mathfrak{M} \subset \mathfrak{B}$ for which $\mathfrak{B} \otimes_\gamma \mathfrak{M}^*$ is not an extension of $\mathfrak{M} \otimes_\gamma \mathfrak{M}^*$.

Proof. The proof is a consequence of Theorem 3.9.

It is not without interest to conclude this discussion with a stronger statement than Corollary 3.4.

THEOREM 3.10. Let $\mathfrak{M}_1$ and $\mathfrak{M}_2$ denote two closed linear manifolds in two Banach spaces $\mathfrak{B}_1$ and $\mathfrak{B}_2$ respectively. Then, there exists a projection P_1 of $\mathfrak{B}_1$ on $\mathfrak{M}_1$ with bound 1, and a projection P_2 of $\mathfrak{B}_2$ on $\mathfrak{M}_2$ with bound 1, if and only if,

(1) $\mathfrak{M}_1 \otimes_\gamma \mathfrak{M}_2 \subset \mathfrak{B}_1 \otimes_\gamma \mathfrak{B}_2$ and

(2) there exists a projection P of $\mathfrak{B}_1 \otimes_\gamma \mathfrak{B}_2$ on $\mathfrak{M}_1 \otimes_\gamma \mathfrak{M}_2$ with bound 1.

Proof. Suppose that P_1 and P_2 exist. Corollary 3.4 tells us that (1) holds. To prove (2) we put

$$P(\Sigma_{i=1}^{m} f_i \otimes g_i) = \Sigma_{i=1}^{m} P_1 f_i \otimes P_2 g_i .$$

It is not difficult to see that P is uniquely defined on $\mathfrak{B}_1 \odot \mathfrak{B}_2$, since $\Sigma_{i=1}^{m} f_i \otimes g_i \simeq \Sigma_{j=1}^{m} f_j' \otimes g_j'$ implies $\Sigma_{i=1}^{m} P_1 f_i \otimes P_2 g_i \simeq \Sigma_{j=1}^{m} P_1 f_j' \otimes P_2 g_j'$ (Lemma 1.6). Clearly, P is additive on $\mathfrak{B}_1 \odot \mathfrak{B}_2$, identical on $\mathfrak{M}_1 \odot \mathfrak{M}_2$; its range is $\mathfrak{M}_1 \odot \mathfrak{M}_2$. Now for any expression $\Sigma_{j=1}^{m} f_j' \otimes g_j' \simeq \Sigma_{i=1}^{m} f_i \otimes g_i$ we have,

$$\gamma(\Sigma_{i=1}^{m} P_1 f_i \otimes P_2 g_i) = \gamma(\Sigma_{j=1}^{m} P_1 f_j' \otimes P_2 g_j') \leq$$
$$\Sigma_{j=1}^{m} \|P_1 f_j'\| \, \|P_2 g_j'\| \leq \Sigma_{j=1}^{m} \|f_j'\| \, \|g_j'\|$$

Thus, Definition 2.5 gives

$$\gamma(P(\Sigma_{i=1}^{m} f_i \otimes g_i)) \leq \gamma(\Sigma_{i=1}^{m} f_i \otimes g_i) .$$

Thus, the bound of P on $\mathfrak{B}_1 \odot_\gamma \mathfrak{B}_2$ is 1. P can be extended in a unique manner to an operator P_0 on $\mathfrak{B}_1 \otimes_\gamma \mathfrak{B}_2$. Clearly, P_0 is a projection of $\mathfrak{B}_1 \otimes_\gamma \mathfrak{B}_2$ on $\mathfrak{M}_1 \otimes_\gamma \mathfrak{M}_2$ with bound 1.

Conversely, we assume that (1) and (2) hold. Choose a $g_0 \in \mathfrak{M}_2$ such that $\|g_0\| = 1$. By [1, p. 55] there exists an additive and bounded functional G on $\mathfrak{M}_2$ such that $G(g_0) = \|g_0\| = 1$, $|||G||| = 1$. Clearly, $G(g)g_0$ is a projection of bound 1 of $\mathfrak{M}_2$ on the linear set of multiples of g_0. Hence, $\mathfrak{M}_1 \otimes_\gamma g_0 \subset \mathfrak{M}_1 \otimes_\gamma \mathfrak{M}_2$ by Lemma 3.7. Now, for $\sum_{i=1}^{m} f_i \otimes g_i$ in $\mathfrak{M}_1 \odot \mathfrak{M}_2$ define

$$Q\left(\sum_{i=1}^{m} f_i \otimes g_i\right) = \sum_{i=1}^{m} G(g_i) f_i \otimes g_0 .$$

We readily verify that Q is additive, invariant under equivalence, assumes identical values on $\mathfrak{M}_1 \otimes_\gamma g_0$ and its range is $\mathfrak{M}_1 \otimes_\gamma g_0$. Furthermore, its bound on $\mathfrak{M}_1 \odot_\gamma \mathfrak{M}_2$ is 1. Thus, it may be extended in a unique manner to a projection Q from $\mathfrak{M}_1 \otimes_\gamma \mathfrak{M}_2$ on $\mathfrak{M}_1 \otimes_\gamma g_0$ without changing its bound. Thus, QP is a projection with bound 1 of $\mathfrak{B}_1 \otimes_\gamma \mathfrak{B}_2$ on $\mathfrak{M}_1 \otimes_\gamma g_0$. Its "contraction" clearly furnishes a projection with bound 1 of $\mathfrak{B}_1 \otimes_\gamma g_0$ on $\mathfrak{M}_1 \otimes_\gamma g_0$. The last two spaces may be identified with $\mathfrak{B}_1$ and $\mathfrak{M}_1$, since their elements are of the form $f \otimes g_0$ with $f \in \mathfrak{B}_1$ or $f \in \mathfrak{M}_1$, and $\gamma(f \otimes g_0) = \|f\| \|g_0\| = \|f\|$. Thus, there exists a projection with bound 1 of $\mathfrak{B}_1$ on $\mathfrak{M}_1$. A similar statement applies to $\mathfrak{B}_2$ and $\mathfrak{M}_2$. This concludes the proof.

By Theorem 3.9, in general, γ is not of "local character". Therefore, a statement for γ analogous to Lemma 2.12 proven for λ, is in general not true.

CHAPTER IV

IDEALS OF OPERATORS

1. Ideals of operators.

For a crossnorm α, the Banach space of all operators A from $\mathfrak{B}_1$ into $\mathfrak{B}_2^*$ (from $\mathfrak{B}_2$ into $\mathfrak{B}_1^*$) of finite α-norm (where $\| A \|_\alpha$ represents the norm of A) may be characterized as $(\mathfrak{B}_1 \otimes_\alpha \mathfrak{B}_2)^*$. In the present section we shall characterize all crossnorms α for which $(\mathfrak{B}_1 \otimes_\alpha \mathfrak{B}_2)^*$ possesses an "ideal" property. We shall see later that the characterization presented here in the language of Banach spaces is equivalent to "unitary invariance" of a crossnorm when applied to Hilbert spaces.

We recall that an operator X on a Banach space $\mathfrak{B}$ determines an "adjoint" operator X^* on $\mathfrak{B}^*$ with $|||X^*||| = |||X|||$. Similarly, X^{**} is determined on $\mathfrak{B}^{**}$; X^{**} is always an extension of X. When $\mathfrak{B}$ is reflexive, $X^{**} = X$, and every operator $\hat{X}$ on $\mathfrak{B}^*$ is the adjoint of some operator on $\mathfrak{B}$, namely $\hat{X} = (\hat{X}^*)^*$.

To simplify our discussion we shall assume that the Banach spaces $\mathfrak{B}_1$, $\mathfrak{B}_2$, under consideration are reflexive. This is equivalent to the statement that $\mathfrak{B}_1^*$ and $\mathfrak{B}_2^*$ are reflexive [10]. It is readily seen that some of the lemmas listed below also hold for perfectly general Banach spaces.

DEFINITION 4.1. A Banach space $\mathfrak{B}$ whose elements are operators A from $\mathfrak{B}_1$ into $\mathfrak{B}_2^*$ and where the norm $\| A \|$ of an operator A is not necessarily equal to its bound will be termed an ideal if it satisfies the following two conditions:

(i) $A \in \mathfrak{B}$ implies $YAX \in \mathfrak{B}$

for any pair of operators X and Y on $\mathfrak{B}_1$ and $\mathfrak{B}_2^*$ respectively.

(ii) $\| YAX \| \leq |||Y||| \, \|A\| \, |||X|||$.

LEMMA 4.1. Let α denote a given crossnorm on $\mathfrak{B}_1 \odot \mathfrak{B}_2$ and $\sum_{i=1}^{m} f_i \otimes g_i$ a fixed expression. Then, there exists an operator A from $\mathfrak{B}_1$ into $\mathfrak{B}_2^*$ of finite α-norm for which

(1) $\|A\|_\alpha = 1$,

(2) $\sum_{i=1}^{m} (Af_i)g_i = \alpha(\sum_{i=1}^{m} f_i \otimes g_i)$.

Proof. Let $\sum_{i=1}^{m} f_i \otimes g_i$ be a fixed expression. By [1, p. 55, Theorem 3] there exists an additive bounded functional $\mathfrak{F}$ on $\mathfrak{B}_1 \otimes_\alpha \mathfrak{B}_2$ with $|||\mathfrak{F}||| = 1$ and $\mathfrak{F}(\sum_{i=1}^{m} f_i \otimes g_i) = \alpha(\sum_{i=1}^{m} f_i \otimes g_i)$. By Theorem 3.1, this $\mathfrak{F}$ generates an operator A of finite α-norm for which $\| A \|_\alpha = |||\mathfrak{F}||| = 1$ and $\sum_{i=1}^{m} (Af_i)g_i = \mathfrak{F}(\sum_{i=1}^{m} f_i \otimes g_i)$ for every expression $\sum_{i=1}^{m} f_i \otimes g_i$. Clearly, A has the desired properties (1) and (2). This concludes the proof.

LEMMA 4.2. Let α denote a uniform crossnorm on $\mathfrak{B}_1 \odot \mathfrak{B}_2$. Then, its associate α' is also uniform on $\mathfrak{B}_1^* \odot \mathfrak{B}_2^*$.

Proof. Let $\hat{S}$ and $\hat{T}$ denote two operators on $\mathfrak{B}_1^*$ and $\mathfrak{B}_2^*$ respectively. Their adjoints S and T are operators on $\mathfrak{B}_1$ and $\mathfrak{B}_2$. For a fixed expression $\sum_{j=1}^{m} F_j \otimes G_j$ in $\mathcal{B}_1^* \odot \mathcal{B}_2^*$ we have,

$$(\sum_{j=1}^{m} \hat{S}F_j \otimes \hat{T}G_j)(\sum_{i=1}^{n} f_i \otimes g_i) =$$

$$(\sum_{j=1}^{m} F_j \otimes G_j)(\sum_{i=1}^{n} Sf_i \otimes Tg_i) \leq$$

$$\alpha'(\sum_{j=1}^{m} F_j \otimes G_j) \ \alpha(\sum_{i=1}^{n} Sf_i \otimes Tg_i)$$

By assumption α is uniform. Recalling that $|||\hat{S}||| = |||S|||$ and $|||\hat{T}||| = |||T|||$ the extreme right of the last inequality is

$$\leq \alpha'(\sum_{j=1}^{m} F_j \otimes G_j) \ |||\hat{S}||| \ |||\hat{T}||| \ \alpha(\sum_{i=1}^{n} f_i \otimes g_i) \ .$$

The last holds for every expression $\sum_{i=1}^{n} f_i \otimes g_i$. This by Definition 2.2 implies

$$\alpha'(\sum_{j=1}^{m} \hat{S}F_j \otimes \hat{T}G_j) \leq |||\hat{S}||| \ |||\hat{T}||| \ \alpha'(\sum_{j=1}^{m} F_j \otimes G_j) \ .$$

This concludes the proof.

THEOREM 4.1. For a crossnorm α on $\mathcal{B}_1 \odot \mathcal{B}_2$, the Banach space of all operators from $\mathcal{B}_1$ into $\mathcal{B}_2^*$ (from $\mathcal{B}_2$ into $\mathcal{B}_1^*$) of finite α-norm forms an ideal if and only if, α is uniform.

Proof. We assume first that α is a uniform crossnorm. Let A denote an operator from $\mathcal{B}_1$ into $\mathcal{B}_2^*$ of finite α-norm, while X and Y stand for any two operators on $\mathcal{B}_1$ and $\mathcal{B}_2^*$ respectively. Then,

$$|\sum_{i=1}^{n} (YAXf_i)g_i| = |\sum_{i=1}^{n} (AXf_i)(Y^*g_i)| \leq \|A\|_\alpha \ \alpha(\sum_{i=1}^{n} Xf_i \otimes Y^*g_i) .$$

Since α is uniform, the extreme right (remembering $|||Y||| = |||Y^*|||$) is

$$\leq \|A\|_\alpha \ |||X||| \ |||Y||| \ \alpha(\sum_{i=1}^{n} f_i \otimes g_i) \ .$$

Since our inequality holds for every expression $\sum_{i=1}^{n} f_i \otimes g_i$, Definition 3.2 gives

$$\|YAX\|_\alpha \leq |||X||| \ |||Y||| \ \|A\|_\alpha \ .$$

Thus, $\|A\|_\alpha < +\infty$ implies $\|YAX\|_\alpha < +\infty$, that is, YAX is of finite

α-norm and satisfies inequality (ii) of Definition 4.1.

To prove the converse assume that the Banach space of operators A from $\mathcal{B}_1$ into $\mathcal{B}_2^*$ of finite α-norm (and where $\|A\|_\alpha$ represents the norm of A) forms an ideal, that is, satisfies conditions (i) and (ii) of Definition 4.1.

We remark first that for a fixed expression $\sum_{i=1}^m f_i \otimes g_i$ all operators A of finite α-norm (Definition 3.2) satisfy the inequality

$$\left| \sum_{i=1}^m (Af_i) g_i \right| \leq \|A\|_\alpha \, \alpha\left(\sum_{i=1}^m f_i \otimes g_i\right) .$$

Furthermore, by Lemma 4.1 the equality sign is acctually assumed by some A with $\|A\|_\alpha = 1$.

Now, let $\sum_{i=1}^m f_i \otimes g_i$ be a fixed expression and X , Y , any two operators on $\mathcal{B}_1$ and $\mathcal{B}_2$ respectively. By the previous remark we can find an operator A^o for which

$$\|A^o\|_\alpha = 1 \quad , \quad \sum_{i=1}^m (A^o X f_i) Y g_i = \alpha\left(\sum_{i=1}^m X f_i \otimes Y g_i\right) .$$

Recalling that by (ii)

$$\|Y^* A^o X\|_\alpha \leq |||Y^*||| \; \|A^o\|_\alpha \; |||X||| = |||Y||| \; |||X|||$$

we have

$$\alpha\left(\sum_{i=1}^m X f_i \otimes Y g_i\right) = \left| \sum_{i=1}^m (Y^* A^o X f_i) g_i \right| \leq$$

$$\|Y^* A X\|_\alpha \, \alpha\left(\sum_{i=1}^m f_i \otimes g_i\right) \leq |||Y||| \; |||X||| \; \alpha\left(\sum_{i=1}^m f_i \otimes g_i\right) .$$

Thus, α is uniform. This concludes the proof.

THEOREM 4.2. For a crossnorm α, $(\mathcal{B}_1 \otimes_\alpha \mathcal{B}_2)^*$ determines an ideal of operators from $\mathcal{B}_1$ into $\mathcal{B}_2^*$ (from $\mathcal{B}_2$ into $\mathcal{B}_1^*$) if and only if, α is uniform.

Proof. The proof is a consequence of Theorem 4.1 and the fact that $(\mathcal{B}_1 \otimes_\alpha \mathcal{B}_2)^*$ represents the space of all operators from $\mathcal{B}_1$ into $\mathcal{B}_2^*$ (from

$\mathcal{B}_2$ into $\mathcal{B}_1^*$) of finite α-norm (as was proven in Theorem 3.1).

THEOREM 4.3. Let α denote a crossnorm. Whenever the Banach space of operators from $\mathcal{B}_1$ into $\mathcal{B}_2^*$ of finite α-norm forms an ideal, then also the Banach space of operators from $\mathcal{B}_2$ into $\mathcal{B}_1^*$ of finite α-norm forms an ideal.

Proof. By Theorem 3.1 and Remark 3.3, $(\mathcal{B}_1 \otimes_\alpha \mathcal{B}_2)^*$ may be interpreted as the space of all operators from $\mathcal{B}_1$ into $\mathcal{B}_2^*$ of finite α-norm as well as the space of all operators from $\mathcal{B}_2$ into $\mathcal{B}_1^*$ of finite α-norm. Thus, an application of Theorem 4.2 concludes the proof.

THEOREM 4.4. Let α be a uniform crossnorm $\geq \lambda$ on $\mathcal{B}_1 \odot \mathcal{B}_2$. Then, $\mathcal{B}_1^* \otimes_{\alpha'} \mathcal{B}_2^*$ that is, the Banach space of all operators from $\mathcal{B}_1$ into $\mathcal{B}_2^*$ (from $\mathcal{B}_2$ into $\mathcal{B}_1^*$) of finite α-norm, approximable in that norm by operators of finite rank forms an ideal. Clearly, it includes all operators of finite rank.

Proof. Let A (an operator from $\mathcal{B}_1$ into $\mathcal{B}_2^*$) be an element of $\mathcal{B}_1^* \otimes_{\alpha'} \mathcal{B}_2^*$ and S, $\hat{T}$, denote any two operators on $\mathcal{B}_1$ and $\mathcal{B}_2^*$ respectively. It is sufficient to verify that $\hat{T}AS$ belongs to $\mathcal{B}_1^* \otimes_{\alpha'} \mathcal{B}_2^*$ since condition (ii) of Definition 4.1 is satisfied even for all operators generated by $(\mathcal{B}_1 \otimes_\alpha \mathcal{B}_2)^*$ (Theorem 4.1 and 4.2). This is easy: We put $\hat{S} = S^*$. The operator A is represented by a sequence of expressions $\sum_{j=1}^{m_p} F_j^{(p)} \otimes G_j^{(p)}$ in $\mathcal{B}_1^* \odot \mathcal{B}_2^*$ fundamental with respect to α'. Since α is uniform, α' is such (Lemma 4.2). This implies that also the sequence $\sum_{j=1}^{m_p} \hat{S}F_j^{(p)} \otimes \hat{T}G_j^{(p)}$ is fundamental relative to the norm α'. It can be readily verified that the last sequence determines $\hat{T}AS$. This concludes the proof.

CHAPTER V

CROSSED UNITARY SPACES

1. Preliminary remarks.

In this chapter we investigate cross-spaces generated by unitary and in particular by Hilbert spaces. To be able to make use of some known theorems for operators on unitary spaces, the following comments are vital:

Let $\mathfrak{H}$ be a complete unitary space, hence a Banach space, and as before $\mathfrak{H}^*$ stand for the space of additive bounded functionals on $\mathfrak{H}$ where the bound represents the norm. Let $f \in \mathfrak{H}$ be fixed. Define f^* by means of the relation:

$$f^*(\varphi) = (\varphi, f) \qquad \text{for} \quad \varphi \in \mathfrak{H} .$$

Then, f^* is an additive bounded functional on $\mathfrak{H}$, with a bound equal to $\|f\|$. Conversely, every additive bounded functional on $\mathfrak{H}$ can be represented in the above fashion uniquely. The one-to-one correspondence $f^* \rightleftarrows f$ between $\mathfrak{H}^*$ and $\mathfrak{H}$ obviously satisfies the following relations:

$$f_1^* \rightleftarrows f_1 \quad \text{and} \quad f_2^* \rightleftarrows f_2 \quad \text{implies} \quad f_1^* + f_2^* \rightleftarrows f_1 + f_2 .$$

However,

$$f^* \rightleftarrows f \quad \text{implies} \quad af^* \rightleftarrows \bar{a}f \quad \text{for any complex number } a .$$

For this reason the correspondence $f^* \rightleftarrows f$ is referred to as a <u>conjugate isomorphism</u>. $\mathfrak{H}^*$ is a Banach space. Moreover $f^* \rightleftarrows f$ implies

$\| f^{*} \| = \| f \|$. Defining however,

$$(f^{*}, g^{*}) = (g , f)$$

we introduce an inner product in $\mathfrak{H}^{*}$, and $\| f^{*} \|$ is the norm that goes with it. In the future we shall denote by $\overline{\mathfrak{H}}$, the space $\mathfrak{H}^{*}$ to which an inner product has been added in the described above manner; its elements will accordingly be denoted by $\bar{f}$, $\bar{g}$, $\bar{\varphi}$, $\bar{\psi}$, ...

Thus, with $\mathfrak{H}$ also $\overline{\mathfrak{H}}$ is a unitary space. Despite the conjugate isomorphism $f^{*} \rightleftarrows f$ between them we shall consider them conceptually distinct.

Now let $\mathfrak{B}_1$ and $\mathfrak{B}_2$ denote two Banach spaces and $\mathfrak{B}_1^{*}$, $\mathfrak{B}_2^{*}$, their conjugate spaces. As customary we form $f \otimes g$ for $f \in \mathfrak{B}_1$, $g \in \mathfrak{B}_2$, and $F \otimes G$ for $F \in \mathfrak{B}_1^{*}$, $G \in \mathfrak{B}_2^{*}$. By Definition 1.3, the number $F(f)G(g)$ represents their inner product $(F \otimes G)(f \otimes g)$. It has been also suggestive -- although not essential -- to interpret $f \otimes g$ as the operator $G(g)f$ <u>from</u> $\mathfrak{B}_2^{*}$ into $\mathfrak{B}_1$. We follow this pattern in the case of unitary spaces. With $\varphi \in \mathfrak{H}$, $\psi \in \mathfrak{H}$, the $\varphi \otimes \psi$ may be interpreted as the operator $(\varphi \otimes \psi)\bar{f} = (\psi , f) \varphi$ <u>from</u> $\overline{\mathfrak{H}}$ into $\mathfrak{H}$. Due to the conjugate isomorphism between $\overline{\mathfrak{H}}$ and $\mathfrak{H}$, the $\varphi \otimes \psi$ may be also interpreted as a conjugate additive operator <u>on</u> $\mathfrak{H}$. Since it is desirable to deal with additive and not conjugate additive operators on $\mathfrak{H}$, we are led to consider $\mathfrak{H} \odot \overline{\mathfrak{H}}$ instead of $\mathfrak{H} \odot \mathfrak{H}$. Accordingly we shall form the inner product for elements $\bar{f} \otimes g$ of $\overline{\mathfrak{H}} \odot \mathfrak{H}$ by elements $\varphi \otimes \bar{\psi}$ of $\mathfrak{H} \odot \overline{\mathfrak{H}}$ conforming to our previous conventions:

$$(\bar{f} \otimes g)(\varphi \otimes \bar{\psi}) = (\varphi , f)(g , \psi) .$$

Both $\bar{f} \otimes g$ and $\varphi \otimes \bar{\psi}$ may be interpreted as operators on $\mathfrak{H}$. The first is clearly an operator on $\overline{\mathfrak{H}}$, defined by

$$(\bar{f} \otimes g)\bar{\varphi} = (g, \varphi)\bar{f} \quad \text{for } \bar{\varphi} \in \overline{\mathfrak{H}}$$

hence also an operator on $\mathfrak{H}$, where

$$(\bar{f} \otimes g)\varphi = (\varphi, g)f \quad \text{for } \varphi \in \mathfrak{H} .$$

The second is an operator from $\overline{\overline{\mathfrak{H}}} = \mathfrak{H}$ into $\mathfrak{H}$, defined by the relation

$$(\varphi \otimes \bar{\psi})f = (f, \psi)\varphi \quad \text{for } f \in \mathfrak{H} .$$

Thus, both $\mathfrak{H} \odot \overline{\mathfrak{H}}$ and $\overline{\mathfrak{H}} \odot \mathfrak{H}$ are isomorphic to the linear space of operators of finite rank.

2. The canonical resolution for operators.

Throughout the present Chapter we shall assume that $\mathfrak{H}$ represents a Hilbert space (unless otherwise specified). All operators considered here are (unless otherwise explicitly stated) assumed to be defined on the whole space $\mathfrak{H}$ and their range contained in $\mathfrak{H}$. To these we shall apply quite often the decomposition given by J. von Neumann in [22], and expressed below in Lemma 5.1.

For our further discussion we recall that for every operator A the operator A^*A is Hermitean and definite. Following [22] or [21, p. 249] there exists a unique Hermitean and definite operator B, such that $B^2 = A^*A$. We write symbolically $B = \text{abs}(A)$ or $B = (A^*A)^{\frac{1}{2}}$. If A is of finite rank, then both A^*A and $\text{abs}(A)$ are also of finite rank.

LEMMA 5.1. Let A denote an operator on $\mathfrak{H}$. There exists a partial isometric operator W whose initial set is the closed linear manifold determined by the range of abs(A) such that, the following formulae hold:

(i) $A = W\,\mathrm{abs}(A)$.

(ii) $\mathrm{abs}(A) = W^*A$.

(iii) $\mathrm{abs}(A^*) = W\,\mathrm{abs}(A)\,W^*$.

(iv) $\mathrm{abs}(A) = \mathrm{abs}(A)\,W^*W$.

The decomposition presented in these formulae is unique in the following sense: $A = W_1 B_1$ where $B_1 \geqslant 0$ and W_1 is partially isometric with the initial set the closure of the range of B_1, implies $B_1 = \mathrm{abs}(A)$ and $W_1 = W$. In the case A is of finite rank we may assume that W is unitary.

Proof. First we prove the existence of the above decomposition. A^*A is Hermitean definite on $\mathfrak{H}$. For $B = \mathrm{abs}(A)$ we have, $B^* = B$ and $B^*B = B^2 = A^*A$. Hence always

$$\| Af \| = \| Bf \| .$$

Putting $C = \mathrm{abs}(A^*) = (AA^*)^{\frac{1}{2}}$ we obtain a similar relationship between C and A^*, namely,

$$\| A^*f \| = \| Cf \| .$$

Thus, it is possible to construct the partial isometric operator W, whose initial and final set are the closed linear manifolds determined by the ranges of B and A respectively, and such that

$$A = WB = CW , \qquad A^* = BW^* = W^*C$$

$$C = WBW^* , \qquad B = W^*CW .$$

This proves the existence of the stated decomposition.

To prove the uniqueness, suppose $A = WB = W_1 B_1$, where $B \geqslant 0$, $B_1 \geqslant 0$ and W as well as W_1 are partial isometric with the initial set being the closed linear manifolds determined by the ranges of B and B_1 respectively. The final set of both W and W_1 is the closed linear manifold determined by the range of A. Then, $A^* = BW^* = B_1 W_1^*$, $A^*A = BW^*WB = B_1 W_1^* W_1 B_1$ that is, $A^*A = B^2 = B_1^2$, hence $B = B_1$ and therefore $W = W_1$.

Suppose finally that A is of finite rank. Since $\| Af \| = \| Bf \|$ the ranges of A and B are of the same dimension. Due to this finite equidimensionality the isometric transformation W' of all Bf on all Af is one-to-one, and possesses a unitary extension (not unique however!). This concludes the proof.

The unique decomposition $A = W \operatorname{abs}(A)$ proven above for operators will be referred to in the future as the "canonical resolution".

3. A characterization of the completely continuous operators.

LEMMA 5.2. Let φ, ψ, denote two elements in $\mathfrak{H}$. We consider $\varphi \otimes \bar{\psi}$ as an operator on $\mathfrak{H}$, whose defining equation is

$$(\varphi \otimes \bar{\psi})f = (f, \psi)\varphi \qquad \text{for } f \in \mathfrak{H}.$$

Then,

(i) $(\varphi \otimes \bar{\psi})^* = \psi \otimes \bar{\varphi}$.

(ii) $a\varphi \otimes \bar{\psi} = a(\varphi \otimes \bar{\psi})$

(ii') $\varphi \otimes \overline{a\psi} = \bar{a}(\varphi \otimes \bar{\psi})$

(a denotes any complex number) .

(iii) $(\varphi_1 + \varphi_2) \otimes \bar{\psi} = \varphi_1 \otimes \bar{\psi} + \varphi_2 \otimes \bar{\psi}$.

(iii') $\varphi \otimes \overline{(\psi_1 + \psi_2)} = \varphi \otimes \overline{\psi_1} + \varphi \otimes \overline{\psi_2}$.

(iv) $(\varphi_1 \otimes \overline{\psi_1})(\varphi_2 \otimes \overline{\psi_2}) = (\varphi_2 , \psi_1)\, \varphi_1 \otimes \overline{\psi_2}$.

(v) $A(\varphi \otimes \overline{\psi}) = A\varphi \otimes \overline{\psi}$

(v') $(\varphi \otimes \overline{\psi})A = \varphi \otimes \overline{A^* \psi}$

(A denotes any operator).

Proof. These relationships are a simple consequence of the definition of $\varphi \otimes \overline{\psi}$ and may be verified immediately by an elementary calculation.

Similarly, the meaning of the symbol $\sum_{i=1}^{m} a_i \varphi_i \otimes \overline{\psi_i}$ is clear.

LEMMA 5.3. Given two nos (φ_i) and (ψ_i) and a sequence of non-negative numbers (a_i) for which $\lim a_i = 0$. Then, the infinite series

$$\sum_i a_i (g , \psi_i)\, \varphi_i$$

is convergent for every g in $\mathfrak{H}$; we denote its sum by Ag . Thus, the sum determines an operator A , which we shall denote symbolically by

$$A = \sum_i a_i \varphi_i \otimes \overline{\psi_i} .$$

Its bound

$$|||A||| = \sup_i a_i .$$

Proof. For $n > m$ we have,

$$\sum_{i=m}^{n} a_i^2 \,|(g , \psi_i)|^2 \leq \sup_{m \leq i \leq n} a_i^2 \|g\|^2 .$$

Thus, Ag is defined for every g [18, Theorem 1.6].

We evaluate $|||A|||$. For $\|g\| = 1$, we have,

$$\|Ag\|^2 = \sum_i a_i^2 \,|(g , \psi_i)|^2 \leq \sup_i a_i^2 .$$

On the other hand putting ψ_i for g we get

$$\|A\psi_i\| = \|a_i\varphi_i\| = a_i \qquad \text{for } i = 1, 2, \ldots\ldots .$$

The last two relationships prove that

$$|||A||| = \sup_{\|g\|=1} \|Ag\| = \sup_i a_i .$$

This concludes the proof.

LEMMA 5.4. An operator A is completely continuous if and only if, $\mathrm{abs}(A)$ is completely continuous.

Proof. The proof is a consequence of Lemma 5.1 (i) and (ii).

In our future discussion we shall avail ourselves of the following characterization of the precise class of completely continuous operators:

LEMMA 5.5. The completely continuous op-rators A on $\mathfrak{H}$ are exactly those which have the "canonical representation" $\sum_i a_i \varphi_i \otimes \overline{\psi}_i$ where (φ_i) and (ψ_i) are orthonormal sets and all the a_i's are positive. Furthermore, $\lim a_i = 0$ in case the sum $\sum_i a_i \varphi_i \otimes \overline{\psi}_i$ has an infinite number of terms. The above representation is unique. The a_i's form the positive point proper values (with multiplicities) of $\mathrm{abs}(A)$.

Proof. Let A be completely continuous. By Lemma 5.4, $\mathrm{abs}(A)$ is also completely continuous. Since it is also Hermitean and definite, (by Hilbert) it possesses a pure point spectrum, that is, there exists a complete orthonormal set ψ_k' of proper vectors of $\mathrm{abs}(A)$. Let a_1' , a_2' ,......... denote the corresponding point proper values (with multiplicities), that is,

$$\mathrm{abs}(A)\psi_k' = a_k'\psi_k' .$$

The definiteness implies $a_k' \geq 0$. The complete continuity implies

$\lim a'_k = 0$. Clearly, $\operatorname{abs}(A) = \sum_k a'_k \psi'_k \otimes \overline{\psi'_k}$. We consider the positive a'_k's. Relabeling the subscripts we get $\operatorname{abs}(A) = \sum_i a_i \psi_i \otimes \overline{\psi_i}$ where all the a_i's are positive, and the (ψ_i) form a nos.

By Lemma 5.1, $A = W\operatorname{abs}(A)$, where W is isometric on the closed linear manifold determined by (ψ_i) . Hence, $(W\psi_i)$ forms a nos. Thus,

$$A = W\left(\sum_i a_i \psi_i \otimes \overline{\psi_i}\right) = \sum_i a_i W\psi_i \otimes \overline{\psi_i} = \sum_i a_i \varphi_i \otimes \overline{\psi_i} .$$

Conversely. Suppose that for a sequence a_1 , a_2, ... of positive numbers we have $\lim a_i = 0$. By Lemma 5.3, for two nos (φ_i) and (ψ_i) , the sum $\sum_i a_i \varphi_i \otimes \overline{\psi_i}$ represents an operator A with a bound equal to $\sup a_i$. Since,

$$\left|\left|\left| A - \sum_{i=1}^{m} a_i \varphi_i \otimes \overline{\psi_i} \right|\right|\right| = \left|\left|\left| \sum_{i>m} a_i \varphi_i \otimes \overline{\psi_i} \right|\right|\right| = \sup_{i>m} a_i ,$$

the operator A can be approximated in bound by a sequence of operators of finite rank. Thus, A must be completely continuous by [1, p. 96] .

Since, $A^*A = \sum_i a_i^2 \psi_i \otimes \overline{\psi_i}$, we have $\operatorname{abs}(A) = \sum_i a_i \psi_i \otimes \overline{\psi_i}$, that is, the a_i's represent the positive point proper values of $\operatorname{abs}(A)$. This concludes the proof.

Quite often completely continuous operators (and therefore also those of finite rank) will be represented in the "canonical form" given by Lemma 5.5.

4. The Schmidt-class of operators.

LEMMA 5.6. Given an operator A on $\mathfrak{K}$, and two cnos (φ_i) and (ψ_j) , the infinite sums

$$\sum_i \|A\varphi_i\|^2 , \quad \sum_j \|A^*\psi_j\|^2 , \quad \sum_{i,j} |(A\varphi_i, \psi_j)|^2$$

of non-negative terms, are either absolutely convergent or properly divergent, at any rate, they all have well-defined values. For these sums the following statement holds: They are equal to each other and independent of (φ_i), (ψ_j).

Proof. Clearly, $(A^*\psi_j, \varphi_i) = \overline{(A\varphi_i, \psi_j)}$. By Parseval's equality $\|A\varphi_i\|^2 = \Sigma_j |(A\varphi_i, \psi_j)|^2$. Hence,

$$\Sigma_i \|A\varphi_i\|^2 = \Sigma_{i,j} |(A\varphi_i, \psi_j)|^2 = \Sigma_{i,j} |(A^*\psi_j, \varphi_i)|^2 = \Sigma_j \|A^*\psi_j\|^2 .$$

This concludes the proof.

DEFINITION 5.1. We denote by $(\sigma(A))^2$ the common value of the three sums in Lemma 5.6. The operators A for which $\sigma(A) < +\infty$ form the E. Schmidt-class (sc).

LEMMA 5.7.

(i) $A \in (sc)$ $A^* \in (sc)$.

(i') $\sigma(A) = \sigma(A^*)$.

(ii) $A \in (sc) \rightarrow aA \in (sc)$ (a is any complex number).

(iii) $A, B \in (sc) \rightarrow (A + B) \in (sc)$.

(iv) $A \in (sc) \rightarrow AX$ and $XA \in (sc)$ (X denotes any operator).

(iv') $\sigma(AX)$ and $\sigma(XA) \leq |||X||| \, \sigma(A)$.

(v) $\varphi \otimes \overline{\psi} \in (sc)$ for φ, ψ in $\mathfrak{H}$.

Proof. (i), (i'), (ii) are obvious.

(iii): $((A + B)\varphi_i, \psi_j) = (A\varphi_i, \psi_j) + (B\varphi_i, \psi_j)$ hence

$$|((A + B)\varphi_i, \psi_j)|^2 \leq 2(|(A\varphi_i, \psi_j)|^2 + |(B\varphi_i, \psi_j)|^2) .$$

Thus, the third sum of Lemma 5.6 yields the desired result.

(iv), (iv'): It suffices to consider XA, since $AX = (X^* A^*)^*$ (remembering (i), (i'), and $|||X||| = |||X^*|||$).

Now, $\| XA\varphi_i\|^2 \leq |||X|||^2 \| A\varphi_i\|^2$. Therefore the first sum of Lemma 5.6 furnishes the desired result.

(v): For a cnos (ψ_i) we have,

$$\Sigma_i \| (\varphi \otimes \bar{\psi})\psi_i\|^2 = \Sigma_i \| (\psi_i , \psi)\varphi\|^2 \leq \Sigma_i |(\psi_i , \psi)|^2 \|\varphi\|^2$$

By Bessel's inequality the extreme right is $\leq \|\psi\|^2 \|\varphi\|^2$.

It is a consequence of (v) and (iii) above that all operators of finite rank are in the Schmidt-class.

LEMMA 5.8. Given two operators A, B in (sc) and a cnos (φ_i), the sum $\Sigma_i (A\varphi_i , B\varphi_i)$ is absolutely convergent and independent of (φ_i).

Proof. Absolute convergence: $|(A\varphi_i , B\varphi_i)| \leq \frac{1}{2}(\| A\varphi_i\|^2 + \| B\varphi_i\|^2)$. The result follows by considering the first sum of Lemma 5.6 and Definition 5.1.

Independence of (φ_i):

$$\mathfrak{R}(A\varphi_i , B\varphi_i) = \frac{1}{4}(\|(A+B)\varphi_i\|^2 - \|(A-B)\varphi_i\|^2).$$

Therefore, $\mathfrak{R}\Sigma_i(A\varphi_i , B\varphi_i) = \frac{1}{4}((\sigma(A+B))^2 - (\sigma(A-B))^2)$, (we use Lemma 5.7 (ii), (iii), and Definition 5.1). Thus, $\mathfrak{R}\Sigma_i(A\varphi_i , B\varphi_i)$ is independent of (φ_i). Replacing A by iA (i stands for the imaginary unit) we see that $\mathfrak{I}\Sigma_i(A\varphi_i , B\varphi_i)$ is independent of (φ_i).

DEFINITION 5.2. For A, B in (sc) the value of the sum of Lemma 5.8 is denoted by (A, B).

LEMMA 5.9.

(i) $(B, A) = \overline{(A, B)}$.

(ii) $(aA, B) = a(A, B)$ ⎫ (a is any complex number).

(ii') $(A, aB) = \bar{a}(A, B)$ ⎭

(iii) $(A_1 + A_2, B) = (A_1, B) + (A_2, B)$.

(iii') $(A, B_1 + B_2) = (A, B_1) + (A, B_2)$.

(iv) $(A, A) \geq 0$.

(iv') $(A, A) = 0$ only for $A = 0$.

(v) $(A^*, B^*) = \overline{(A, B)}$.

(vi) $(XA, B) = (A, X^*B)$ ⎫ (X denotes any operator).

(vi') $(AX, B) = (A, BX^*)$ ⎭

Proof. (i)--(iii') are obviously a consequence of Definition 5.2.

(iv): Obvious, since $(A\varphi_i, A\varphi_i) = \|A\varphi_i\|^2$.

(iv'): Using the first sum of Lemma 5.6, $\sigma(A) = 0$ means $A\varphi_i = 0$ for all cnos (φ_i) , that is, $A\varphi = 0$ for all $\|\varphi\| = 1$. This means $A = 0$.

(v): By (iv) this becomes $\sigma(A^*) = \sigma(A)$ for $A = B$. The last equation holds by Lemma 5.7 (i'). Replacing in this particular case $A = B$, A by $A + B$ or by $A - B$ and subtracting, we obtain $\mathfrak{R}(A^*, B^*) = \mathfrak{R}(A, B)$. Next, replacing A by iA we obtain $\mathfrak{I}(A^*, B^*) = -\mathfrak{I}(A, B)$. Hence, $(A^*, B^*) = \overline{(A, B)}$.

(vi): Obvious, since $(XA\varphi_i, B\varphi_i) = (A\varphi_i, X^*B\varphi_i)$.

(vi'): Replacing in (vi) A , B , X by A^* , B^* , X^* , and applying (v) to both sides of the equality, we obtain the desired result.

REMARK 5.1. Lemma 5.7 (ii) and (iii) state that the operators in (sc) form a linear space. Lemma 5.9 (i)--(iv') makes it clear that (A , B) is an inner product in (sc) and

$$(A, A)^{\frac{1}{2}} = \sigma(A)$$

the norm that goes with it. Therefore, we also have Schwarz's inequality:

$$|(A, B)| \leq \sigma(A)\, \sigma(B) .$$

REMARK 5.2. In the particular case when both A and B are operators of finite rank hence in the Schmidt-class, say $A = \sum_{i=1}^{m} \varphi_i \otimes \overline{\psi_i}$ and $B = \sum_{j=1}^{m} \overline{f_j} \otimes g_j$, then the inner product (A , B) defined above, coincides with $(\sum_{j=1}^{m} \overline{f_j} \otimes g_j)(\sum_{i=1}^{m} \varphi_i \otimes \overline{\psi_i})$ -- the one stated -- in Definition 1.3. To see this let (ω_k) denote a cnos in $\mathfrak{K}$. Then,

$$(A, B) = \sum_k \left(\sum_{i=1}^{m} (\omega_k, \psi_i)\varphi_i , \sum_{j=1}^{m} (\omega_k, g_j) f_j\right) =$$

$$\sum_{j=1}^{m} \sum_{i=1}^{m} (\varphi_i, f_j) \sum_k (\omega_k, \psi_i)\overline{(\omega_k, g_j)} .$$

By Parseval's identity [18, p. 10]:

$$\sum_k (\omega_k, \psi_i)\overline{(\omega_k, g_j)} = (g_j, \psi_i) .$$

Thus,

$$(A, B) = \sum_{j=1}^{m} \sum_{i=1}^{m} (\varphi_i, f_j)(g_j, \psi_i) .$$

This concludes the proof.

In conclusion of this section let us add that some properties of the Schmidt-class of operators have been investigated in [20] , and later in [15] and [16] .

5. The trace-class of operators.

LEMMA 5.10. Given an operator A such that $A = C^*B$ where B and C are in (sc), and a cnos (φ_i), the sum $\sum_i (A\varphi_i, \varphi_i)$ is absolutely convergent, independent of (φ_i), and equal to (B, C).

Proof. This is a simple consequence of Lemma 5.8 and Definition 5.2.

REMARK 5.3. Lemma 5.10 also shows that (B, C) (for B, C, in (sc)) depends on C^*B only, and not on B, C, separately.

REMARK 5.4. Lemma 5.10 deals with operators of the form $A = C^*B$ where B, C, are in (sc). Considering Lemma 5.7 (i), and replacing B, C, by C, B^*, it is evident that the operators A considered in Lemma 5.10 may be characterized in the form $A = BC$ with B, C, in (sc).

DEFINITION 5.3. The operators A of Lemma 5.10 (or of Remark 5.4) form the trace-class (tc). For such an A we denote the value of the sum of Lemma 5.10 by $t(A)$, and term the "trace" of A.

Consider an operator A. Then A^*A is a Hermitean definite operator. By [22, p. 302], $\text{abs}(A) = (A^*A)^{\frac{1}{2}}$ is a uniquely defined definite operator. By the same argument this is also true for $(\text{abs}(A))^{\frac{1}{2}}$. Since $\text{abs}(A)$ is definite, for a given cnos (φ_i), the sum $\sum_i (\text{abs}(A)\varphi_i, \varphi_i)$ has ≥ 0 terms, hence it is either absolutely convergent or properly divergent, at any rate it has a well-defined value.

The cnos (φ_i) in Lemma 5.11 (iv) below, should be viewed as arbitrarily given but fixed, i.e., the criterion is valid if applied with one cnos (φ_i) only, and it does not matter how this (φ_i) is chosen.

LEMMA 5.11. The following statements are equivalent to each other:

(i) $A \in (tc)$.

(ii) $abs(A) \in (tc)$.

(iii) $(abs(A))^{\frac{1}{2}} \in (sc)$.

(iv) $\sum_i (abs(A)\varphi_i, \varphi_i) < +\infty$.

Proof. We shall show that (iii)→(i)→(ii)→(iv)→(iii).

(iii)→(i): Let $(abs(A))^{\frac{1}{2}} \in (sc)$. By Lemma 5.7 (iv), $W(abs(A))^{\frac{1}{2}} \in (sc)$. Hence, $W(abs(A))^{\frac{1}{2}}(abs(A))^{\frac{1}{2}} = Wabs(A) = A \in (t$

(i)→(ii): Let $A = BC$, where B , C are in (sc) . Then, $abs(A) = W^*A = W^*BC$, and $W^*B \in (sc)$, by Lemma 5.7 (iv). This proves $abs(A) \in (tc)$.

(ii)→(iv): This is immediate by Definition 5.3 and Lemma 5.10.

(iv)→(iii): $\|(abs(A))^{\frac{1}{2}}\varphi_i\|^2 = (abs(A)\varphi_i, \varphi_i)$, hence $\sum_i \|(abs(A))^{\frac{1}{2}}\varphi_i\|^2 < +\infty$. Thus, the first sum of Lemma 5.6 yields $(abs(A))^{\frac{1}{2}} \in (sc)$. This concludes the proof.

LEMMA 5.12. Let a denote a complex number and X any operator.

(i) $A \in (tc) \rightleftarrows A^* \in (tc)$.

(ii) $A \in (tc) \rightarrow aA \in (tc)$.

(iii) A , $B \in (tc) \rightarrow A + B \in (tc)$.

(iv) $A \in (tc) \rightarrow AX \in (tc)$ and $XA \in (tc)$.

Proof. (i): Obvious by Lemma 5.7 (i), since $A = BC$ is equivalent to $A^* = C^*B^*$.

(ii): Clear, by Lemma 5.7 (ii).

(iv): Immediate, by Lemma 5.7 (iv), since $A = BC$ implies $AX = B(CX)$ and $XA = (XB)C$.

(iii): By Lemma 5.1, $\mathrm{abs}(A + B) = W^*(A + B)$. Thus, by Lemma 5.11 (iv), it suffices to establish the absolute convergence of

$$\Sigma_i (\mathrm{abs}(A + B)\varphi_i, \varphi_i) = \Sigma_i (W^*(A + B)\varphi_i, \varphi_i) =$$
$$\Sigma_i ((W^*A\varphi_i, \varphi_i) + (W^*B\varphi_i, \varphi_i)) .$$

Now, A, B are in (tc). Therefore by (iv) above, W^*A, W^*B are in (tc). This implies that the sums $\Sigma_i (W^*A\varphi_i, \varphi_i)$, $\Sigma_i (W^*B\varphi_i, \varphi_i)$ are absolutely convergent and therefore $\Sigma_i(\mathrm{abs}(A + B)\varphi_i, \varphi_i)$ is also absolutely convergent. This concludes the proof.

LEMMA 5.13. In (i)--(iii) both A, B are in (tc); in (iv) either A or B is in (tc). The expressions in (i)--(iii) are defined by Lemma 5.12 (i)--(iii); the expressions in (iv) are defined by Lemma 5.12 (iv).

(i) $\overline{t(A)} = t(A^*)$.

(ii) $t(aA) = at(A)$ (a denotes any complex number).

(iii) $t(A + B) = t(A) + t(B)$.

(iv) $t(AB) = t(BA)$.

Proof. (i)--(iii): Obvious, by Lemma 5.10 and Definition 5.3.

(iv): Assume by symmetry, that $A \in$ (tc) and write X for B. Following Lemma 5.10 we put $A = C^*B$, where B, C are in (sc).

By Definition 5.3,

$$t(AX) = t(C^*BX) = (BX, C) .$$

$$t(XA) = t(X C^*B) = (B, CX^*) .$$

The two extreme right-hand expressions are equal to each other by Lemma 5.9 (vi') .

DEFINITION 5.4. For $A \in (tc)$, we define $m(A) = t(abs(A))$; the last number is defined by Lemma 5.11 (ii) .

LEMMA 5.14. In what follows A , B are in (tc) , and X denotes any operator.

(i) $m(A^*) = m(A)$.

(ii) $m(aA) = |a| m(A)$.

(iii) $m(A + B) \leq m(A) + m(B)$.

(iv) $m(A) \geq 0$; $= 0$ holds for $A = 0$ only.

(v) $m(AX)$ and $m(XA) \leq |||X||| m(A)$.

(vi) $|t(A)| \leq m(A)$.

Proof. (i): $m(A^*) = t(abs(A^*)) = t(W abs(A) W^*)$ by Lemma 5.1 (iii). The extreme right of the last equality may be written as $t(abs(A)W^*W)$ by Lemma 5.13 (iv). Thus, by Lemma 5.1 (iv), $m(A^*) = m(A)$.

(ii): Obvious, since $(aA)^*(aA) = |a|^2 A^*A$. Therefore,

$$abs(aA) = |a| \; abs(A) .$$

(v): It suffices to consider XA , since $AX = (X^*A^*)^*$, (we use (i) and $|||X^*||| = |||X|||$). By Lemma 5.1 (i),

$$abs(XA) = W_1^*XA , \qquad A = W abs(A)$$

where $|||W||| = 1$ and $|||W_1^*||| = 1$. Hence, $\mathrm{abs}(XA) = Y\mathrm{abs}(A)$ with $Y = W_1^* X W$. Therefore, $|||Y||| \leq |||X|||$. Now, by Lemma 5.11 (iii), $(\mathrm{abs}(A))^{\frac{1}{2}} \in (sc)$. Consequently, Lemma 5.1 (iv) gives $Y(\mathrm{abs}(A))^{\frac{1}{2}} \in (sc)$. Therefore,

$$m(XA) = t(\mathrm{abs}(XA)) = t(Y\mathrm{abs}(A)) =$$
$$t(Y(\mathrm{abs}(A))^{\frac{1}{2}}(\mathrm{abs}(A))^{\frac{1}{2}}) =$$
$$\left((\mathrm{abs}(A))^{\frac{1}{2}},\ (Y(\mathrm{abs}(A))^{\frac{1}{2}})^*\right).$$

Using Schwarz's inequality (Remark 5.1) we get

$$m(XA) \leq \sigma((\mathrm{abs}(A))^{\frac{1}{2}})\,\sigma(Y(\mathrm{abs}(A))^{\frac{1}{2}}).$$

Thus, by Lemma 5.7 (i') and (iv')

$$m(XA) \leq |||Y|||\left(\sigma((\mathrm{abs}(A))^{\frac{1}{2}})\right)^2 \leq$$
$$|||X|||\left((\mathrm{abs}(A))^{\frac{1}{2}},\ (\mathrm{abs}(A))^{\frac{1}{2}}\right) =$$
$$|||X|||\,t((\mathrm{abs}(A))^{\frac{1}{2}}(\mathrm{abs}(A))^{\frac{1}{2}}) =$$
$$|||X|||\,t(\mathrm{abs}(A)) = |||X|||\,m(A).$$

(vi): By Lemma 5.1 (i),

$$t(A) = t(W\mathrm{abs}(A)) = t(W(\mathrm{abs}(A))^{\frac{1}{2}}(\mathrm{abs}(A))^{\frac{1}{2}}) =$$
$$\left((\mathrm{abs}(A))^{\frac{1}{2}},\ (W(\mathrm{abs}(A))^{\frac{1}{2}})^*\right).$$

Applying Schwarz's inequality to the extreme right we get,

$$|t(A)| \leq \sigma((\mathrm{abs}(A))^{\frac{1}{2}})\,\sigma((W(\mathrm{abs}(A))^{\frac{1}{2}})^*).$$

Recalling that $|||W^*||| = 1$ and using Lemma 5.7 (i') and (iv') we have,

$$|t(A)| \leq \left(\sigma((\mathrm{abs}(A))^{\frac{1}{2}})\right)^2 =$$
$$\left((\mathrm{abs}(A))^{\frac{1}{2}},\ (\mathrm{abs}(A))^{\frac{1}{2}}\right) =$$
$$t\left((\mathrm{abs}(A))^{\frac{1}{2}}(\mathrm{abs}(A))^{\frac{1}{2}}\right) =$$
$$t(\mathrm{abs}(A)) = m(A).$$

(iii): Formulae (i) and (ii) of Lemma 5.1 furnish

$$A = W\,\mathrm{abs}(A) \;,\quad B = W_1\,\mathrm{abs}(B) \;,\quad \mathrm{abs}(A+B) = W_2(A+B)$$

where $|||W||| = 1$, $|||W_1||| = 1$, $|||W_2||| = 1$. Therefore,

$\mathrm{abs}(A+B) = X\mathrm{abs}(A) + Y\mathrm{abs}(B)$ where $X = W_2W$, $Y = W_2W_1$; $|||X||| \leqslant 1$, $|||Y||| \leqslant 1$. Now,

$$m(A+B) = t(\mathrm{abs}(A+B)) = t(X\mathrm{abs}(A) + Y\mathrm{abs}(B)) =$$

$$t(X\mathrm{abs}(A)) + t(Y\mathrm{abs}(B)) .$$

Applying Schwarz's inequality and (i'), (iv) of Lemma 5.7 or (vi), (v) of this Lemma, we see that the extreme right of the last equality is not greater than

$$|||X|||\, m(A) + |||Y|||\, m(B) \leqslant m(A) + m(B) .$$

(iv): Clearly,

$$m(A) = t(\mathrm{abs}(A)) = t((\mathrm{abs}(A))^{\frac{1}{2}} (\mathrm{abs}(A))^{\frac{1}{2}}) =$$

$$((\mathrm{abs}(A))^{\frac{1}{2}} ,\; (\mathrm{abs}(A))^{\frac{1}{2}}) = (\sigma((\mathrm{abs}(A))^{\frac{1}{2}}))^2 .$$

Therefore, $m(A) \geqslant 0$, and $m(A) = 0$ implies $\sigma((\mathrm{abs}(A))^{\frac{1}{2}}) = 0$. By Lemma 5.9 (iv') it follows $(\mathrm{abs}(A))^{\frac{1}{2}} = 0$, that is, $(\mathrm{abs}(A))^2 = 0$. This implies $A^*A = 0$, and therefore $A = 0$.

LEMMA 5.15.

(i) $\varphi \otimes \bar{\psi} \in (tc)$.

(ii) $t(\varphi \otimes \bar{\psi}) = (\varphi, \psi)$.

(iii) $m(\varphi \otimes \bar{\psi}) = \|\varphi\|\,\|\psi\|$.

Proof. We may assume that $\varphi \neq 0$ and $\psi \neq 0$ otherwise the proof is trivial. By Lemma 5.1 (ii) and (v),

$$(\text{abs}(\varphi \otimes \bar{\psi}))^2 = (\varphi \otimes \bar{\psi})^*(\varphi \otimes \bar{\psi}) =$$
$$(\psi \otimes \bar{\varphi})(\varphi \otimes \bar{\psi}) = \|\varphi\|^2 \psi \otimes \bar{\psi} .$$

Now, put $\varphi_1 = \frac{\psi}{\|\psi\|}$. Then $\|\varphi_1\| = 1$, and φ_1 is a proper vector of the last operator corresponding to the proper value $\|\varphi\|^2\|\psi\|^2$. Further, every f which is orthogonal to φ_1 , that is, to ψ , is a proper vector of the last operator corresponding to the proper value 0 . We extend φ_1 to a cnos (φ_i) . Then, $\varphi_1, \varphi_2, \varphi_3, \ldots$ are proper vectors of $(\text{abs}(\varphi \otimes \bar{\psi}))^2$ corresponding to the proper values $\|\varphi\|^2\|\psi\|^2$, 0 , 0 ,.... . Therefore, $\varphi_1, \varphi_2, \varphi_3, \ldots$ are also proper vectors of $\text{abs}(\varphi \otimes \bar{\psi})$ corresponding to the proper values $\|\varphi\|\|\psi\|$, 0 , 0 ,.... . Consequently,

$$((\text{abs}(\varphi \otimes \bar{\psi}))\varphi_i, \varphi_i) = \begin{cases} \|\varphi\|\|\psi\| & \text{for } i = 1 \\ 0 & \text{for } i > 1 \end{cases} .$$

We have,

$$\Sigma_i((\text{abs}(\varphi \otimes \bar{\psi}))\varphi_i, \varphi_i) = \|\varphi\|\|\psi\| < +\infty .$$

Thus, by Lemma 5.6, $\text{abs}(\varphi \otimes \bar{\psi}) \in (\text{tc})$ and $\varphi \otimes \bar{\psi} \in (\text{tc})$. Moreover, $m(\varphi \otimes \bar{\psi}) = \|\varphi\|\|\psi\|$. This proves (i) and (iii).

To prove (ii), we take any cnos (ψ_i) . Using Parseval's equation,

$$t(\varphi \otimes \bar{\psi}) = \Sigma_i((\varphi \otimes \bar{\psi})\psi_i, \psi_i) = \Sigma_i(\psi_i, \psi)(\varphi, \psi_i) =$$
$$\Sigma_i(\varphi, \psi_i)(\psi, \psi_i) = (\varphi, \psi) .$$

This proves (ii). This concludes the proof.

REMARK 5.5. It is a consequence of Lemma 5.12 (ii) and (iii), that the operators of (tc) form a linear set. Because of Lemma 5.15 (i), (tc) contains all operators of finite rank. Lemma 5.11 proves that A in (tc) is equivalent to abs(A) in (tc) ; therefore m(A) is defined. Lemma 5.14 (ii), (iii), (iv), prove that m(A) is a norm on (tc) , in fact a crossnorm (Lemma 5.15).

6. Symmetric gauge functions.

In this section we present some elementary results for symmetric gauge functions which we shall make use of in the future. These will permit us to derive later an explicit representation for all uniform crossnorms on $\mathfrak{H} \otimes \mathfrak{K}$ as well as some relationships between them. First we investigate gauge functions on a finite dimensional space.

DEFINITION 5.5. A real function $\Phi(u_1, \ldots, u_n)$ on the n-dimensional space of n-tuples $(u_1, \ldots, u_n)$ of real numbers is termed a gauge function if it satisfies the following conditions:

(i) $\Phi(u_1, \ldots, u_n) > 0$ unless $u_1 = \ldots = u_n = 0$.

(ii) $\Phi(cu_1, \ldots, cu_n) = |c|\, \Phi(u_1, \ldots, u_n)$ for any constant c .

(iii) $\Phi(u_1 + u_1', \ldots, u_n + u_n') \leq \Phi(u_1, \ldots, u_n) + \Phi(u_1', \ldots, u_n')$.

Φ will be termed symmetric if in addition to (i), (ii), (iii), it satisfies the following condition:

(iv) $\Phi(u_1, \ldots, u_n) = \Phi(\varepsilon_1 u_{1'}, \ldots, \varepsilon_n u_{n'})$

where $\varepsilon_i = \pm 1$ and $1', \ldots, n'$ denotes any permutation on $1, \ldots, n$.

To simplify our formulae we shall always assume that Φ also satisfies the following condition:

(v) $\Phi(1, 0, \ldots, 0) = 1$.

A symmetric gauge function is non-decreasing in each variable u_i, that is, $|u_i| \leqslant |u_i'|$ for $i = 1, \ldots, n$ implies $\Phi(u_1, \ldots, u_n) \leqslant \Phi(u_1', \ldots, u_n')$. This is precisely the content of the following Lemma:

LEMMA 5.16. Let $\Phi(u_1, \ldots, u_n)$ denote a symmetric gauge function on $\mathcal{R}_n$. Then, for $0 \leqslant p_i \leqslant 1$ we have,

$$\Phi(p_1 u_1, \ldots, p_n u_n) \leqslant \Phi(u_1, \ldots, u_n) .$$

Proof. By virtue of (iv) we may suppose that all the u_i's are $\geqslant 0$. It is clear by induction that it is sufficient to establish the last relation when $p_i \neq 1$ occurs only for one i, that is,

$$\Phi(u_1, \ldots, u_{i-1}, pu_i, u_{i+1}, \ldots, u_n) \leqslant \Phi(u_1, \ldots, u_{i-1}, u_i, u_{i+1}, \ldots, u_n)$$

for $0 \leqslant p < 1$. The last assertion follows from the following simple direct calculation:

$$\Phi(u_1, \ldots, pu_i, \ldots, u_n) =$$

$$\Phi\left(\frac{1+p}{2}u_1 + \frac{1-p}{2}u_1, \ldots, \frac{1+p}{2}u_i + \frac{1-p}{2}(-u_i), \ldots, \frac{1+p}{2}u_n + \frac{1-p}{2}u_n\right) \leqslant$$

$$\Phi\left(\frac{1+p}{2}u_1, \ldots, \frac{1+p}{2}u_i, \ldots, \frac{1+p}{2}u_n\right) + \Phi\left(\frac{1-p}{2}u_1, \ldots, \frac{1-p}{2}(-u_i), \ldots \frac{1-p}{2}u_n\right)$$

$$\frac{1+p}{2}\Phi(u_1, \ldots, u_i, \ldots, u_n) + \frac{1-p}{2}\Phi(u_1, \ldots, (-u_i), \ldots, u_n) =$$

$$\frac{1+p}{2}\Phi(u_1, \ldots, u_i, \ldots, u_n) + \frac{1-p}{2}\Phi(u_1, \ldots, u_i, \ldots, u_n)$$

$$\Phi(u_1, \ldots, u_i, \ldots, u_n) .$$

REMARK 5.6. In the proof of our Lemma we have not assumed at all that Φ satisfies condition (v) of Definition 5.5.

LEMMA 5.17. Let $\Phi(u_1,\ldots,u_n)$ denote a symmetric gauge function on $\mathfrak{R}_n$. Then,

$$\max_i |u_i| \leqslant \Phi(u_1,\ldots,u_n) .$$

Proof. Lemma 5.16 gives,

$$\Phi(0,\ldots,0,u_i,0,\ldots,0) \leqslant \Phi(u_1,\ldots,u_{i-1},u_i,u_{i+1},\ldots,u_n) .$$

By (v) and (ii) of Definition 5.5 for Φ the left side of the last inequality equals $|u_i|$, that is,

$$|u_i| \leqslant \Phi(u_1,\ldots,u_i,\ldots,u_n) \qquad \text{for} \quad i = 1,\ldots,n .$$

This concludes the proof.

LEMMA 5.18. For a symmetric gauge function $\Phi(u_1,\ldots,u_n)$ on $\mathfrak{R}_n$ we have,

$$\Phi(u_1,\ldots,u_n) \leqslant \sum_{i=1}^{n} |u_i| .$$

Proof. The proof is a simple consequence of conditions (iii), (iv) and (v) for Φ.

LEMMA 5.19. A gauge function $\Phi(u_1,\ldots,u_n)$ is continuous.

Proof. Conditions (ii) and (iii) for Φ furnish

$$|\Phi(u_1',\ldots,u_n') - \Phi(u_1,\ldots,u_n)| \leqslant$$
$$|\Phi(u_1-u_1',\ldots,u_n-u_n')| .$$

The last is $\leqslant \sum_{i=1}^{n} |u_i - u_i'|$ by Lemma 5.18. This concludes the proof.

We are about to define an associate of a given symmetric gauge function $\Phi(u_1,\ldots,u_n)$ on $\mathfrak{R}_n$. For a fixed n-tuple $(v_1,\ldots,v_n)$,

$$\text{(a)} \qquad \frac{u_1 v_1 + \ldots\ldots + u_n v_n}{\Phi(u_1, \ldots\ldots, u_n)}$$

represents a continuous function on the compact (that is, closed and bounded) set of n-tuples $(u_1, \ldots\ldots, u_n)$ for which $|u_1| + \ldots. + |u_n| = 1$. Hence it assumes a maximum which we shall denote by $\Psi(v_1, \ldots\ldots, v_n)$. The last of course may be also defined as the maximum of the numbers (a) over the set of all n-tuples $(u_1, \ldots\ldots, u_n) \neq (0, \ldots\ldots, 0)$.

The proof of the following two Lemmas is immediate:

LEMMA 5.20. $\Psi(v_1, \ldots\ldots, v_n)$ is a gauge function whenever $\Phi(u_1, \ldots\ldots, u_n)$ is such. If Φ is symmetric, the same holds for Ψ.

LEMMA 5.21. $u_1 v_1 + \ldots. + u_n v_n \leq \Phi(u_1, \ldots\ldots, u_n)\, \Psi(v_1, \ldots\ldots, v_n)$.

The n-dimensional space of n-tuples $(u_1, \ldots\ldots, u_n)$ on which there is defined a gauge function $\Phi(u_1, \ldots\ldots, u_n)$ will be denoted by $\mathfrak{R}_n(\Phi)$.

LEMMA 5.22. Every additive functional F on $\mathfrak{R}_n(\Phi)$ determines an n-tuple of numbers $(v_1, \ldots\ldots, v_n)$ such that,

$$(1) \qquad F(u_1, \ldots\ldots, u_n) = u_1 v_1 + \ldots. + u_n v_n \quad \text{for} \quad (u_1, \ldots\ldots, u_n) \quad \text{in} \quad \mathfrak{R}_n(\Phi$$

Furthermore, its bound

$$(2) \qquad |||F||| = \max \frac{u_1 v_1 + \ldots. + u_n v_n}{\Phi(u_1, \ldots\ldots, u_n)},$$

where the last max is extended over all n-tuples $(u_1, \ldots\ldots, u_n) \neq (0, \ldots\ldots, 0)$

Converselv. Given an n-tuple of numbers $(v_1, \ldots\ldots, v_n)$, then relatior ship (1) above, determines an additive functional F. Its bound is given by (2

Proof. Let F be an additive functional on $\mathfrak{L}_m(\Phi)$. Put $F(1,0,\ldots,0) = v_1, \ldots, F(0,0,\ldots,1) = v_m$. Then, (1) above is a consequence of additivity for F. Clearly, (2) follows from (1) and the definition of a bound for an additive functional. That the converse holds, is immediate. This concludes the proof.

LEMMA 5.23. The conjugate space of $\mathfrak{L}_m(\Phi)$ may be characterized as $\mathfrak{L}_m(\Psi)$.

Proof. This is a consequence of Lemma 5.22 and the definition of Ψ for a given Φ.

DEFINITION 5.6. $\Psi(v_1, \ldots, v_m)$ is termed the associate gauge function of $\Phi(u_1, \ldots, u_m)$.

LEMMA 5.24. Any gauge function $\Phi(u_1, \ldots, u_m)$ on $\mathfrak{L}_m$ is at the same time the associate of its associate gauge function $\Psi(v_1, \ldots, v_m)$.

Proof. By Lemma 5.23, the conjugate space of $\mathfrak{L}_m(\Phi)$ may be characterized as $\mathfrak{L}_m(\Psi)$. But every finite dimensional Banach space is reflexive. Therefore, the conjugate of $\mathfrak{L}_m(\Psi)$ may be characterized as $\mathfrak{L}_m(\Phi)$. This concludes the proof.

Consider the set whose elements are infinite sequences of real numbers $(u_1, u_2, \ldots)$ having only a finite number of non-zero terms. Defining addition of elements and multiplication of an element by a scalar in the obvious fashion we obtain a linear set $\mathfrak{L}$.

DEFINITION 5.5'. Under a symmetric gauge function Φ on $\mathcal{R}$ we shall understand any function $\Phi(u_1, u_2, \ldots)$ defined on $\mathcal{R}$ subject to conditions (i) -- (v) of Definition 5.5 in which the n-tuples are replaced by elements of $\mathcal{R}$, that is, infinite sequences having only a finite number of non-zero terms.

Clearly, the set of all elements $(u_1, \ldots, u_n, 0, 0, \ldots)$ of $\mathcal{R}$ with $u_i = 0$ for $i > n$, may be identified with $\mathcal{R}_n$. A given gauge function $\Phi(u_1, u_2, \ldots)$ on $\mathcal{R}$ defines a gauge function $\Phi_n(u_1, \ldots, u_n) = \Phi(u_1, \ldots, u_n, 0, 0, \ldots)$ on $\mathcal{R}_n$ for $n = 1, 2, \ldots$. Each $\Phi_n(u_1, \ldots, u_n)$ determines on $\mathcal{R}_n$ an associate $\Psi_n(v_1, \ldots, v_n)$.

LEMMA 5.25. For a given n-tuple $(v_1, \ldots, v_n)$ we have,

$$\Psi_n(v_1, \ldots, v_n) = \Psi_{n+1}(v_1, \ldots, v_n, 0) = \Psi_{n+2}(v_1, \ldots, v_n, 0, 0) = \ldots$$

Their common value we shall denote by $\Psi(v_1, \ldots, v_n, 0, 0, \ldots)$.

Proof. By definition, we have,

$$\Psi_n(v_1, \ldots, v_n) = \max_{(u_1, \ldots, u_n)} \frac{u_1 v_1 + \ldots + u_n v_n}{\Phi_n(u_1, \ldots, u_n)} =$$

$$\max_{(u_1, \ldots, u_n)} \frac{u_1 v_1 + \ldots + u_n v_n}{\Phi_{n+1}(u_1, \ldots, u_n, 0)} \leq$$

$$\max_{(u_1, \ldots, u_n, u_{n+1})} \frac{u_1 v_1 + \ldots + u_n v_n + u_{n+1} 0}{\Phi_{n+1}(u_1, \ldots, u_n, u_{n+1})} =$$

$$\Psi_{n+1}(v_1, \ldots, v_n, 0) .$$

To prove the converse inequality, we remark by Lemma 5.16, that for any n + values $u_1, \ldots, u_n, u_{n+1}$, we have,

$$\Phi_n(u_1, \ldots, u_n) = \Phi_{n+1}(u_1, \ldots, u_n, 0) \leq \Phi_{n+1}(u_1, \ldots, u_n, u_{n+1}) .$$

Thus,

$$\frac{u_1 v_1 + \ldots + u_n v_n + u_{n+1} 0}{\Phi_{n+1}(u_1, \ldots, u_n, u_{n+1})} \leq \frac{u_1 v_1 + \ldots + u_n v_n}{\Phi_n(u_1, \ldots, u_n)} .$$

Definition 5.6 implies $\Psi_{n+1}(v_1, \ldots, v_n, 0) \leq \Psi_n(v_1, \ldots, v_n)$. Thus,

$$\Psi_{n+1}(v_1, \ldots, v_n, 0) = \Psi_n(v_1, \ldots, v_n) .$$

The rest of the proof is clear. This concludes the proof.

LEMMA 5.26. For a given symmetric gauge function $\Phi(u_1, u_2, \ldots)$ on $\mathfrak{L}$, the $\Psi(v_1, v_2, \ldots)$ is also a symmetric gauge function on $\mathfrak{L}$.

Proof. Clearly, Ψ satisfies properties (i), (ii), (iv) and (v). Thus it is sufficient to verify (iii). Given two sequences representing elements of $\mathfrak{L}$, there exists a number N such that for both sequences the terms with the subscripts $i > N$ are equal to zero. This means, the sequences may be written in the form

$$v_1, \ldots, v_N, 0, 0, \ldots \quad \text{and} \quad v_1', \ldots, v_N', 0, 0, \ldots .$$

We have,

$$\Psi(v_1 + v_1', \ldots, v_N + v_N', 0, 0, \ldots) = \Psi_N(v_1 + v_1', \ldots, v_N + v_N') \leq$$
$$\Psi_N(v_1, \ldots, v_N) + \Psi_N(v_1', \ldots, v_N') =$$
$$\Psi(v_1, \ldots, v_N, 0, 0, \ldots) + \Psi(v_1', \ldots, v_N', 0, 0, \ldots) .$$

This concludes the proof.

DEFINITION 5.6'. We term $\Psi(v_1, v_2, \ldots)$ the associate gauge function of $\Phi(u_1, u_2, \ldots)$ on $\mathfrak{L}$.

LEMMA 5.27. Any symmetric gauge function $\Phi(u_1, u_2, \ldots)$ on $\mathfrak{L}$ is at the same time the associate of its associate $\Psi(v_1, v_2, \ldots)$.

Proof. Denote by $\widetilde{\Psi}$ the associate of Ψ. For $(u_1, u_2, \ldots)$ in $\mathfrak{R}$ there exists an N such that $u_i = 0$ for $i > N$. By definition,

$$\widetilde{\Psi}(u_1, \ldots, u_N, 0, 0, \ldots) = \widetilde{\Psi}_N(u_1, \ldots, u_N) .$$

The last represents the value of the associate of Ψ_N for $(u_1, \ldots, u_N)$. By Lemma 5.24 it must be equal to

$$\Phi_N(u_1, \ldots, u_N) = \Phi(u_1, \ldots, u_N, 0, 0, \ldots) .$$

This concludes the proof.

Once a norm Φ is defined on $\mathfrak{R}$ we obtain a normed linear space $\mathfrak{R}(\Phi)$ which in general will not be cómplete. Details about the possible methods of "completing" $\mathfrak{R}(\Phi)$, that is, imbedding it in a Banach space may be found in a separate publication of the author. Here let us just outline two methods for which the resulting two Banach spaces will -- in the light of the next chapter -- be closely connected with, and shed additional light on, the mutual relationship of the associate and conjugate space for $\mathfrak{H} \otimes_\alpha \overline{\mathfrak{H}}$, whenever α is a unitarily invariant crossnorm:

(1) the Cantor-Meray method [6, p. 106], by considering the fundamental (Cauchy) sequences of elements in $\mathfrak{R}(\Phi)$ and introducing some standard identifications,

(2) the "strong" method, by considering all those infinite sequences of real numbers $(u_1, u_2, \ldots)$ -- having perhaps an infinite number of non-zer terms -- for which

$$\lim_{n \to \infty} \Phi(u_1, \ldots, u_n, 0, 0, \ldots) < +\infty .$$

The last definition makes sense in the light of Lemma 5.16, which clearly carries over to symmetric gauge functions Φ on $\mathfrak{R}$.

The Cantor-Meray "completion" represents the smallest Banach space in which $\mathfrak{R}(\Phi)$ can be imbedded and is clearly always included in the "strong" completion. It is true that in many important cases procedures (1) and (2) furnish the same space. This happens for instance when $p > 1$ and

$$\Phi(u_1, u_2, \ldots) = (\sum_i |u_i|^p)^{\frac{1}{p}} \quad \text{for} \quad (u_1, u_2, \ldots) \text{ in } \mathfrak{R},$$

when both (1) and (2) furnish the space l_p [1, p. 12] . However, in the "limiting" case for $p = \infty$ that is,

$$\Phi(u_1, u_2, \ldots) = \max_i |u_i| ,$$

procedure (1) furnishes the Banach space (c_0) of all sequences converging towards 0 [1, p. 101] , while procedure (2) furnishes the Banach space (m) of all bounded sequences [1, p. 11] .

Clearly, procedure (1) and (2) will furnish the same completed space if and only if, for every sequence of numbers $(u_1, u_2, \ldots)$ for which

$$\lim_{n \to \infty} \Phi(u_1, \ldots, u_n, 0, 0, \ldots) < +\infty$$

we have,

$$\lim_{\substack{m, n \to \infty \\ m > n}} \Phi(u_n, \ldots, u_m, 0, 0, \ldots) = 0 .$$

It is not without interest to conclude this section on symmetric gauge functions with the following simple theorem whose details may be found in the author's publication [17a] :

The conjugate space of the Cantor-Meray closure of $\mathfrak{R}(\Phi)$ may be characterized as the strong closure of $\mathfrak{R}(\Psi)$. Conversely, the strong closure of $\mathfrak{R}(\Phi)$ may be interpreted as the conjugate space of the Cantor-Meray closure of $\mathfrak{R}(\Psi)$.

7. The class of unitarily invariant crossnorms on $\mathfrak{H} \odot \bar{\mathfrak{H}}$.

In the following discussion we shall derive an explicit representation for all unitarily invariant crossnorms α on $\mathfrak{H} \odot \bar{\mathfrak{H}}$, where $\mathfrak{H}$ is any unitary space, that is either a finite dimensional Euclidean space or a Hilbert space. We shall also prove that the class of unitarily invariant crossnorms coincides with the class of uniform crossnorms. Moreover, for these crossnorms $\alpha'' = \alpha$.

In section 1 of this chapter it was pointed out that $\mathfrak{H} \odot \bar{\mathfrak{H}}$ may be represented as the linear set of all operators A on $\mathfrak{H}$ of finite rank. We shall use this representation most of the time and denote the crossnorm α accordingly $\alpha(A)$. For this reason we find it advisable to restate the definition of a norm and later its associate in terms of operators of finite rank:

DEFINITION 5.7. A norm α is any function $\alpha(A)$ of operators A of finite rank satisfying the following conditions:

(i) $\alpha(A) \geqq 0$; $\alpha(A) = 0 \rightarrow A = 0$.

(ii) $\alpha(cA) = |c| \, \alpha(A)$ for any constant c .

(iii) $\alpha(A + B) \leqq \alpha(A) + \alpha(B)$.

A norm is a crossnorm if in addition to (i), (ii), (iii) it also satisfies the following condition:

(iv) $\alpha(A) = |||A|||$ for all operators A of rank 1.

Uniformity for α clearly means that it satisfies:

(v) $\alpha(SAT) \leqq |||S||| \, |||T||| \, \alpha(A)$ for any pair of operators S, T .

A crossnorm is termed unitarily invariant if it satisfies the following condition:

(v') $\alpha(UAV^*) = \alpha(A)$, for any pair of unitary operators U, V

Let α be a given fixed unitarily invariant crossnorm. Consider an operator A of finite rank. Then, A^*A is of finite rank, Hermitean and definite. Hence, abs(A) which is also of finite rank, Hermitean and definite, has a pure point spectrum; all its proper values are $\geqq 0$, and only a finite number of its proper values (with multiplicities) are $\neq 0$. Let $a_1 \geqq a_2 \geqq \ldots \geqq$ stand for these proper values. By Lemma 5.1, $A = U\mathrm{abs}(A)$ where U is unitary. The unitary invariance of α implies $\alpha(A) = \alpha(\mathrm{abs}(A))$.

Suppose that for two operators A, $\bar{A}$ of finite rank the proper values with multiplicities of abs(A) and abs($\bar{A}$) are the same. Then there exists a unitary V such that $V\mathrm{abs}(A)V^* = \mathrm{abs}(\bar{A})$. Hence, $\alpha(\mathrm{abs}(A)) = \alpha(\mathrm{abs}(\bar{A}))$ and therefore $\alpha(A) = \alpha(\bar{A})$. Thus, $\alpha(A)$ depends only on the $a_1, a_2, \ldots$. We define accordingly

$$\Phi(a_1, a_2, \ldots) = \alpha(A),$$

where $a_1 \geqq a_2 \geqq \ldots \geqq 0$ are the point proper values with multiplicities of abs(A).

Thus a unitarily invariant crossnorm α defines a function $\Phi(a_1, a_2, \ldots)$ for all sequences of real numbers $a_1, a_2, \ldots$ which possess these properties:

(1) $a_i \neq 0$ for only a finite number of i's

(2) $a_1 \geqq a_2 \geqq \ldots \geqq 0$.

We extend the function Φ defining it for all sequences for which we assume only (1), as follows: Let $a_1, a_2, \ldots$ satisfy (1), and $\tilde{a}_1, \tilde{a}_2, \ldots$ be that permutation of $|a_1|$, $|a_2|$, ... for which $\tilde{a}_1 \geq \tilde{a}_2 \geq \ldots \geq 0$. Then, $\tilde{a}_1, \tilde{a}_2, \ldots$ satisfies (1) and (2). By definition we put

$$\Phi(a_1, a_2, \ldots) = \Phi(\tilde{a}_1, \tilde{a}_2, \ldots) .$$

The fact that Φ is generated by a unitarily invariant crossnorm implies that it must satisfy some relationships which can be obtained as follows:

Let $\varphi_1, \varphi_2, \ldots$ be a fixed cnos. We define $P_i = \varphi_i \otimes \bar{\varphi}_i$. Let $a_1, a_2, \ldots$ be any sequence of real numbers satisfying (1) and $A = \Sigma_i a_i P_i$. Then, $A^* = \Sigma_i a_i P_i$. Furthermore, $A^*A = \Sigma_i a_i^2 P_i$ and $\mathrm{abs}(A) = \Sigma_i |a_i| P_i$. Hence, $|a_1|, |a_2|, \ldots$ that is, $\tilde{a}_1, \tilde{a}_2, \ldots$ are the point proper values (with multiplicities) of $\mathrm{abs}(A)$. Consequently, $\alpha(A) = \Phi(\tilde{a}_1, \tilde{a}_2, \ldots)$ i.e., $\alpha(A) = \Phi(a_1, a_2, \ldots)$. Thus,

$$\alpha(\Sigma_i a_i P_i) = \Phi(a_1, a_2, \ldots) .$$

Applying the last equation to $a_1, a_2, \ldots$, to $ca_1, ca_2, \ldots$ to $b_1, b_2, \ldots$ and to $a_1 + b_1, a_2 + b_2, \ldots$ respectively, always with the same cnos (φ_i) we ge

(i) $\Phi(a_1, a_2, \ldots) > 0$ unless $a_1 = a_2 = \ldots = 0$.

(ii) $\Phi(ca_1, ca_2, \ldots) = |c| \Phi(a_1, a_2, \ldots)$ for any constant c

(iii) $\Phi(a_1 + b_1, a_2 + b_2, \ldots) \leq \Phi(a_1, a_2, \ldots) + \Phi(b_1, b_2, \ldots$

The unitary invariance of implies clearly,

(iv) $\Phi(a_1, a_2, \ldots) = \Phi(\varepsilon_1 a_{1'}, \varepsilon_2 a_{2'}, \ldots)$

where $|\varepsilon_i| = 1$ and $1', 2', \ldots$ denotes any permutation on the natural numbers.

Finally, since α is a crossnorm, $\alpha(P_1) = 1$. Hence,

(v) $\Phi(1, 0, 0, \ldots) = 1$.

The preceding discussion may be summed up as follows:

LEMMA 5.28. A unitarily invariant crossnorm α on $\mathfrak{K} \odot \overline{\mathfrak{K}}$ generates a symmetric gauge function $\Phi(a_1, a_2, \ldots)$ satisfying conditions (i) -- (v) of Definition 5.5' in the following manner:

We choose a cnos (φ_i). For any sequence of real numbers $a_1, a_2, \ldots$ satisfying (1) define

$$\Phi(a_1, a_2, \ldots) = \alpha(\Sigma_i a_i P_i) .$$

For the so defined Φ we have,

$$\Phi(a_1, a_2, \ldots) = \alpha(A) ,$$

whenever $a_1 \geqq a_2 \geqq \ldots \geqq 0$ represents the point proper values (with multiplicities) of abs(A).

Conversely, we shall prove that every gauge-function satisfying conditions (i) -- (v) of Definition 5.5' can be derived in the indicated manner from a unitarily invariant crossnorm. To achieve this we shall consider first the finite dimensional case. From this, as we shall see later, the proof can be readily extended to the infinite dimensional case.

Our discussion will be based on the following Lemma which may be found in [23, p. 293, Theorem I]. The symbol t(A) stands for the trace of A (Definition 5.3):

LEMMA 5.29. The maximum of $\mathfrak{R}\, t(UAVB)$, where A, B are fixed and U, V, are running over all unitary operators on an n-dimensional unitary space $\mathfrak{K}_n$ is given by

$$\Sigma_{i=1}^{n} a_i b_i$$

where $a_1, \ldots, a_n$ are the proper values (with multiplicities) of abs(A) while $b_1, \ldots, b_n$ denote the proper values (with multiplicities) of abs(B), both monotonously ordered, that is, $a_1 \geqq \ldots \geqq a_n$ and $b_1 \geqq \ldots \geqq b_n$.

Let Φ be a symmetric gauge function on the n-dimensional linear space $\mathcal{R}_n$ of n-tuples of real numbers $(u_1, \ldots, u_n)$ and Ψ stand for its associate. For our immediate application we may assume that Φ satisfies only (i), (ii), (iv) and (v) of Definition 5.5. For any operator X on the n-dimensional unitary space $\mathcal{R}_n$, we put

$$\Phi(X) = \Phi(x_1, \ldots, x_n)$$

where $x_1, \ldots, x_n$ denote the proper values (with multiplicities) of abs(X).

We shall investigate the sup of the numbers $\mathcal{R}t(XA)$ when A is fixed and X is running over all operators for which $\Phi(X) = 1$. Since for unitary U, V, the operators abs(UXV) and abs(X) posses the same proper values, we have, $\Phi(UXV) = \Phi(X)$. Consequently,

$$\sup_{\Phi(X)=1} \mathcal{R}t(XA) = \sup_{\Phi(X)=1} \left(\max_{U,V} \mathcal{R}t(UXVA) \right).$$

By Lemma 5.29, the last equals to

$$\sup \sum_{i=1}^{n} a_i x_i ,$$

where the $a_1 \geqq \ldots \geqq a_n$ are the fixed proper values of abs(A) and the $x_1, \ldots, x_n$ are subject to the restrictions $x_1 \geqq \ldots\ x_n \geqq 0$ and $\Phi(x_1, \ldots, x_n) = 1$. A little consideration shows however that the requirement $x_1 \geqq \ldots \geqq x_n$ may be omitted. In fact, whenever $x_i < x_j$ for a pair $i < j$, then interchanging x_i and x_j we do not affect $\Phi(x_1, \ldots, x_n) = 1$ and do not decrease $\sum_{i=1}^{n} a_i x_i$, the change being $(a_i x_j + a_j x_i) - (a_i x_i + a_j x_j) = (a_i - a_j)(x_j - x_i) \geqq 0$. Thus, the condition $x_1 \geqq \ldots \geqq x_n \geqq 0$ may be replaced by $x_i \geqq 0$. But even the last may be omitted, because whenever $x_i < 0$, then replacing x_i by $-x_i$ we do not affect $\Phi(x_1, \ldots, x_n) = 1$ and as is readily seen do not decrease $\sum_{i=1}^{n} a_i x_i$. Thus, we are really dealing with $\sup \sum_{i=1}^{n} a_i x_i$ where the $x_1, \ldots, x_n$ are subject to the sole restriction $\Phi(x_1, \ldots, x_n) = 1$. By Definition 5.6 the last sup equals $\Psi(a_1, \ldots, a_n)$.

We may sum up our discussion in the following:

LEMMA 5.30. Let $\Phi(u_1, \ldots, u_n)$ be a symmetric gauge function and $\Psi(v_1, \ldots, v_n)$ its associate. Then, $\mathcal{R}\, t(XA)$ when A is fixed and X runs through all operators (on the n-dimensional unitary space $\mathfrak{R}_n$) with $\Phi(X) = 1$, assumes a maximum, and its numerical value is $\Psi(A)$.

We shall use the last Lemma in the following equivalent form:

LEMMA 5.30'. Let $\Phi(u_1, \ldots, u_n)$ be a symmetric gauge function and $\Psi(v_1, \ldots, v_n)$ its associate. Then,

$$\sup |t(XA)| \qquad \text{or} \qquad \sup |(A, X)|$$

when A is fixed and X is assumed to vary over all operators (on the n-dimensional unitary space $\mathfrak{R}_n$) with $\Phi(X) = 1$, assumes a maximum; its numerical value is $\Psi(A)$.

Proof. That both last sup's are equal, follows from the fact that $t(XA) = (A, X^*)$ and $\Phi(X) = \Phi(X^*)$. To prove that $\sup \mathcal{R}\, t(XA) = \sup |t(XA)|$ we argue as follows: Let θ denote any complex number with $|\theta| = 1$. Replacing X by θX, $\Phi(\theta X) = \Phi(X)$ remains unaltered, while $\mathcal{R}(\theta\, t(XA))$ -- as θ varies -- assumes a maximum equal to $|t(XA)|$. The rest is clear. This concludes the proof.

LEMMA 5.31. Let $\Phi(u_1, \ldots, u_n)$ be a symmetric gauge function (Definition 5.5). For an operator A on an n-dimensional unitary space R_n put

$$\Phi(A) = \Phi(a_1, \ldots, a_n) ,$$

where $a_1, \ldots, a_n$ denote the proper values (with multiplicities) of $\mathrm{abs}(A)$. Then,

(i) $\Phi(A) \geqslant 0$; $\Phi(A) = 0$ implies $A = 0$.

(ii) $\Phi(cA) = |c|\,\Phi(A)$ for any constant c .

(iii) $\Phi(A+B) \leqslant \Phi(A) + \Phi(B)$.

(iv) $\Phi(UAV^*) = \Phi(A)$ for unitary U, V .

(v) $\Phi(A) = |||A|||$, when A is of rank $\leqslant 1$.

Proof. (i). That $\Phi(A) \geqslant 0$, is clear. Now, $0 = \Phi(A) = \Phi(a_1, \ldots, a_n)$ implies $a_1 = \ldots = a_n = 0$, that is, $A^*A = 0$ and $A = 0$

(ii). Since $(cA)^*(cA) = |c|^2 A^*A$, the numbers $|c|a_1, \ldots, |c|a_n$ represent the proper values of $\mathrm{abs}(cA)$. Hence,

$$\Phi(cA) = \Phi(|c|a_1, \ldots, |c|a_n) = |c|\,\Phi(a_1, \ldots, a_n) = |c|\,\Phi(A) .$$

(iii). Let $\Psi(v_1, \ldots, v_n)$ be the associate of $\Phi(u_1, \ldots, u_n)$ (Definition 5.6). By Lemma 5.24, $\Phi(u_1, \ldots, u_n)$ is also the associate of $\Psi(v_1, \ldots, v_n)$. By Lemma 5.30',

$$\Phi(A+B) = \sup_{\Psi(X)=1} |(A+B, X)| \leqslant$$

$$\sup_{\Psi(X)=1} |(A, X)| + \sup_{\Psi(X)=1} |(B, X)| = \Phi(A) + \Phi(B) .$$

(iv). This is true since $(UAV^*)^*(UAV^*) = VA^*AV^*$ and A^*A have the same proper values.

(v). If A is of rank 0, the proof is trivial. We assume $A = \varphi \otimes \bar{\psi}$ with $\varphi \neq 0$, $\psi \neq 0$. Then, $(\varphi \otimes \bar{\psi})^*(\varphi \otimes \bar{\psi}) = \|\varphi\|^2 \psi \otimes \bar{\psi}$. Thus,

$$\mathrm{abs}(\varphi \otimes \bar{\psi}) = \frac{\|\varphi\|}{\|\psi\|} \psi \otimes \bar{\psi}$$

and its only positive proper value is $\|\varphi\|\|\psi\|$. Therefore,

$$\Phi(\varphi \otimes \bar{\psi}) = \Phi(\|\varphi\|\|\psi\|, 0, 0, \ldots) = \|\varphi\|\,\|\psi\|$$

which clearly represents the bound of $A = \varphi \otimes \bar{\psi}$. This concludes the proof.

REMARK 5.7. By Lemma 5.31, $\Phi(A)$ is a unitarily invariant cross-norm on $\mathfrak{K}_n \odot \bar{\mathfrak{K}}_n$. The interesting part is naturally the triangle inequality for Φ, that is, $\Phi(A + B) \leq \Phi(A) + \Phi(B)$. Since this is equivalent to saying that $\Phi(A) \leqslant 1$, $\Phi(B) \leqslant 1$; $p, q \geqslant 0$, $p + q = 1$, implies $\Phi(pA + qB) \leqslant 1$, we may express the following:

Let $p \geqslant 0$, $q \geqslant 0$, $p + q = 1$, and let $a_1, a_2, \ldots, a_n$; $b_1, b_2, \ldots, b_n$ and $c_1, c_2, \ldots, c_n$, denote the point proper values (with multiplicities) of $\mathrm{abs}(A)$, $\mathrm{abs}(B)$, and $\mathrm{abs}(pA + qB)$ respectively. Then, for any symmetric gague function Φ satisfying conditions (i)--(iv) of Definition 5.5, the inequalities $\Phi(a_1, \ldots, a_n) \leq 1$, $\Phi(b_1, \ldots, b_n) \leqslant 1$, imply $\Phi(c_1, \ldots, c_n) \leqslant 1$.

We are just ready to extend our result to operators on a Hilbert space and prove the desired converse of Lemma 5.28.

LEMMA 5.32. Let $\Phi(a_1, a_2, \ldots)$ denote a symmetric gauge function on $\mathfrak{K}$ satisfying conditions (i)--(v) of Definition 5.5'. Then the equality

$$\alpha(A) = \Phi(a_1, a_2, \ldots)$$

where $a_1 \geqslant a_2 \geqslant \ldots \geqslant 0$ denote the point proper values (with multiplicities) of $\mathrm{abs}(A)$ defines a unitarily invariant crossnorm α on $\mathfrak{K} \odot \bar{\mathfrak{K}}$.

Proof. Clearly, $\alpha(A)$ is defined for all operators of finite rank on $\mathfrak{R}$. Furthermore, by Lemma 5.31, whenever $\mathfrak{H}$ is any fixed finite dimensional subspace of $\mathfrak{R}$, then $\alpha(A)$ is a norm if A is restricted to those operators for which the ranges of both A and A^* are in $\mathfrak{H}$. Now, given any two operators $\bar{A}$, $\bar{B}$, of finite rank, let $\mathfrak{H}$ be spanned by the ranges of $\bar{A}$, $\bar{A}^*$, $\bar{B}$, $\bar{B}^*$. Clearly, $\mathfrak{H}$ is finite dimensional and $\alpha(A)$ is a norm for a family of operators which contain both (arbitrarily given!) operators $\bar{A}$, $\bar{B}$. The defining properties of a norm never involve more than two operators at a time. This implies that the defined above α is a norm on the set of all operators on $\mathfrak{R}$ of finite rank.

Condition (iv) for Φ implies the unitary invariance of α, and finally (v) for Φ assures for α the "cross-property". This concludes the proof.

In the theorem which follows, as well as throughout the rest of this Chapter, let $\mathfrak{R}$ stand for an n-dimensional unitary space or a Hilbert space. Accordingly, let $\mathfrak{K}$ stand for the set of n-tuples of real numbers, or for the linear space of infinite sequences of real numbers having only a finite number of non-zero terms.

THEOREM 5.1. The class of unitarily invariant crossnorms on $\mathfrak{R} \odot \bar{\mathfrak{R}}$ and the class of symmetric gauge functions generate each other.

Proof. The proof follows from Lemmas 5.28 and 5.32.

The discussion which follows throws additional light on Theorem 5.1 by proving that unitarily invariant and uniform crossnorms coincides.

LEMMA 5.33. Let A denote an operator of finite rank and X be any operator on $\mathfrak{H}$. Let $a_1 \geqq a_2 \geqq \ldots\ldots \geqq 0$ and $b_1 \geqq b_2 \geqq \ldots\ldots \geqq 0$ denote the point proper values (with multiplicities) of abs(A) and abs(XA) respectively. Then, $b_i \leqq |||X||| \, a_i$.

Proof. We use a well-known theorem of Courant [2, p. 27]. It is stated there for finite dimensional Euclidean spaces, but it is clear that it applies equally to Hilbert spaces, always assuming that the operator K in question has a pure point spectrum (with multiplicities) $c_1 \geqq c_2 \geqq \ldots\ldots \geqq 0$. This theorem states:

$$c_i = \min \left(\max \; (Kf, f) \right)$$

where the right side in the last and following three equalities should be read as follows: Let $\mathfrak{M}$ be an $i-1$ dimensional linear manifold. Define $m(\mathfrak{M})$ as the max (Kf, f) over the set of all f's with $\|f\| = 1$ and f orthogonal to $\mathfrak{M}$. We consider then the min of all numbers $m(\mathfrak{M})$ for all $i-1$ dimensional $\mathfrak{M}$.

Now A^*A has a pure point spectrum (with multiplicities) $a_1^2 \geqq a_2^2 \geqq \ldots\ldots$ Hence,

$$a_i^2 = \min \left(\max \; (A^*Af, f) \right).$$

Since, $a_i \geqq 0$ and $(A^*Af, f) = \|Af\|^2$, we have,

$$a_i = \min \left(\max \; \|Af\| \right).$$

Similarly, we form the point proper values (with multiplicities) $b_1 \geqq b_2 \geqq \ldots\ldots$ of abs(XA). Then,

$$b_i = \min \left(\max \; \|XAf\| \right).$$

Clearly, $\| XAf \| \leq |||X||| \, \| Af \|$. Hence, $b_i \leq |||X||| \, a_i$. This concludes the proof.

LEMMA 5.34. Let α be a unitarily invariant crossnorm, and A an operator of finite rank. Then, $\alpha(A) = \alpha(A^*)$.

Proof. By Lemma 5.1, there exists a unitary U such that $A = U\text{abs}(A$ and therefore $A^* = \text{abs}(A)U^*$. Thus, $\alpha(A) = \alpha(\text{abs}(A))$ and $\alpha(A^*) = \alpha(\text{abs}(A))$. This concludes the proof.

LEMMA 5.35. A unitarily invariant crossnorm α is uniform.

Proof. Let A be an operator of finite rank and X any operator. Let $a_1 \geq a_2 \geq \ldots \geq 0$ and $b_1 \geq b_2 \geq \ldots \geq 0$ denote the point proper values (with multiplicities) of $\text{abs}(A)$ and $\text{abs}(XA)$ respectively. By Lemma 5.33, $b_i \leq |||X||| \, a_i$. Hence, $b_i = |||X||| \, p_i a_i$ where $0 \leq p_i \leq 1$. For α we form the symmetric gauge function Φ as stated in Lemma 5.28. Then,

$$\alpha(XA) = \Phi(b_1, b_2, \ldots) = \Phi(|||X||| \, p_1 a_1, |||X||| \, p_2 a_2, \ldots) =$$
$$|||X||| \, \Phi(p_1 a_1, p_2 a_2, \ldots) \leq |||X||| \, \Phi(a_1, a_2, \ldots) = |||X||| \, \alpha(A) .$$

The above inequality sign is a consequence of Lemma 5.16 which clearly carries over to symmetric gauge functions on $\mathfrak{R}$. Thus,

$$\alpha(XA) \leq |||X||| \, \alpha(A) .$$

Now, by Lemma 5.10, $\alpha(A) = \alpha(A^*)$. Thus, by what we have shown already for any operator Y , we have:

$$\alpha(AY) = \alpha((AY)^*) = \alpha(Y^*A^*) \leq$$
$$|||Y^*|||\ \alpha(A^*) = |||Y|||\ \alpha(A) .$$

Therefore,

$$\alpha(YAX) \leq |||Y|||\ \alpha(AX) \leq |||X|||\ |||Y|||\ \alpha(A) .$$

This concludes the proof.

LEMMA 5.36. A uniform crossnorm α is unitarily invariant.

Proof. Let U and V denote any two unitary operators. Then,

$$|||U||| = |||U^*||| = 1 \quad ; \quad |||V||| = |||V^*||| = 1 .$$

We have,

$$\alpha(UAV^*) \leq |||U|||\ \alpha(A)\ |||V^*||| = \alpha(A) .$$

On the other hand

$$\alpha(A) = \alpha(U^*UAV^*V) \leq$$
$$|||U^*|||\ \alpha(UAV^*)\ |||V||| = \alpha(UAV^*) .$$

Thus,

$$\alpha(UAV^*) = \alpha(A) .$$

This concludes the proof.

We combine Lemmata 5.28, 5.32, 5.35 and 5.36 in the following statement:

THEOREM 5.2. The class of uniform crossnorms on $\mathfrak{K} \odot \overline{\mathfrak{K}}$, coincides with the class of unitarily invariant crossnorms on $\mathfrak{K} \odot \overline{\mathfrak{K}}$. The last class and the class of all symmetric gauge functions $\Phi(a_1, a_2, \ldots)$ satisfying conditions (i)--(v) generate each other.

THEOREM 5.3. For a given crossnorm α, the Banach space of all operators on $\mathfrak{K}$ of finite α-norm, that is, $(\mathfrak{K} \otimes_{\alpha} \bar{\mathfrak{K}})^*$ forms an ideal (Definition 4.1) if, and only if, α is unitarily invariant.

Proof. This is a consequence of Theorems 4.2 and 5.2.

THEOREM 5.4. The bound $\lambda(A)$ of an operator A represents the least unitarily invariant, consequently uniform crossnorm.

Proof. Let α be a unitarily invariant crossnorm. By Lemma 5.28, form a corresponding Φ. By Lemma 5.5, an operator A of finite rank may be represented uniquely in the canonical form $\sum_{i=1}^{n} a_i \varphi_i \otimes \bar{\psi}_i$. Its bound

$$\lambda(A) = |||A||| = \max_{1 \leq i \leq n} a_i$$

by Lemma 5.3. On the other hand, Lemma 5.17 gives,

$$\max_{1 \leq i \leq n} a_i \leq \Phi(a_1, a_2, \ldots \quad) = \alpha(A) .$$

Thus, for every unitarily invariant crossnorm α we have $\alpha \geq \lambda$. But λ is clearly unitarily invariant. This concludes the proof.

Although the unitarily invariant (uniform) crossnorms form the significant class of norms it is not without interest to construct examples of crossnorms which are not uniform. From Theorem 5.4 it follows that any crossnorm which is not $\geq \lambda$ is not uniform. We shall construct such crossnorms in the latter part of this chapter.

We conclude this section with the proof that for every unitarily invariant crossnorm α, we have, $\alpha'' = \alpha$.

DEFINITION 5.8. Let $\alpha(X)$ be a unitarily invariant crossnorm on $\mathfrak{R} \odot \overline{\mathfrak{R}}$. For an operator A on $\mathfrak{R}$ of finite rank, let $\alpha'(A)$ be the least constant c satisfying the inequality

$$|(A, X)| \leqslant c\, \alpha(X)$$

for all operators X on $\mathfrak{R}$ of finite rank (or what amounts to the same thing, for all operators X on $\mathfrak{R}$ of finite rank with $\alpha(X) = 1$).

It is a consequence of Remark 5.2, that $\alpha'(A)$ as stated above for operators A on $\mathfrak{R}$ of finite rank, coincides with Definition 2.2 of the associate α' of α, when applied to $\mathfrak{R} \odot \overline{\mathfrak{R}}$.

LEMMA 5.37. Let α be a unitarily invariant crossnorm on $\mathfrak{R} \odot \overline{\mathfrak{R}}$ and Φ the symmetric gauge function it generates (By Lemma 5.28). Then, its associate α' generates the associate gauge function Ψ (Definition 5.6').

Proof. Lemma 5.30' proves this when $\mathfrak{R}$ is a finite dimensional Euclidean space. Here we extend the proof to the case when $\mathfrak{R}$ is a Hilbert space. We show that for an operator A on $\mathfrak{R}$ of finite rank, we have,

$$\alpha'(A) = \Psi(A) = \Psi(a_1, a_2, \ldots),$$

$a_1, a_2, \ldots$ being the point proper values (with multiplicities) of abs(A).

So let A be an operator of finite rank on $\mathfrak{R}$ and $\mathfrak{M}$ the finite say m-dimensional linear manifold generated by the ranges of A and A^*. An operator on $\mathfrak{M}$ may be by Lemma 5.5, represented uniquely in the canonical form $\sum_{i=1}^{m} a_i \varphi_i \otimes \overline{\psi_i}$, which clearly also defines uniquely an operator on $\mathfrak{R}$. Thus, the linear space of operators on $\mathfrak{M}$ may be identified with the linear space of all those operators A on $\mathfrak{H}$ whose range together with the range of A^* is included in $\mathfrak{M}$.

We put $\Phi_m(x_1, \ldots, x_m) = \Phi(x_1, \ldots, x_m, 0, 0, \ldots)$ and let Ψ_m denote the associate of Φ_m. Then clearly, $\Phi(X) = \Phi_m(X)$ and $\Psi_m(X) = \Psi(X)$ for operators X on $\mathfrak{M}$. By Lemma 5.30', $\Psi_m(A)$ is the least constant c satisfying the inequality $|(A, X)| \leq c\,\Phi_m(X)$ for all operators X on $\mathfrak{M}$, hence in the light of the introduced identification, also the inequality $|(A, X)| \leq c\,\Phi(X)$ for all operators X on $\mathfrak{H}$ which may be identified with the operators on $\mathfrak{M}$. Thus, the requirement for this constant $\Psi_m(A)$ is obviously weaker then the one for $\alpha'(A)$ in Definition 5.1 Consequently,

$$\alpha'(A) \geq \Psi_m(A) = \Psi(A) .$$

On the other hand for an operator X on $\mathfrak{H}$ of finite rank we can always find a finite say $m+k$ -dimensional linear manifold containing the ranges of A, A^*, X, X^*. Consequently,

$$\frac{|(A, X)|}{\Phi(X)} = \frac{|(A, X)|}{\Phi_{m+k}(X)} \leq \Psi_{m+k}(A) = \Psi_m(A) = \Psi(A) .$$

Such an inequality holds for any X on $\mathfrak{H}$ of finite rank. This furnishes the converse, $\alpha'(A) \leq \Psi(A)$. This concludes the proof.

THEOREM 5.5. Every unitarily invariant crossnorm α on $\mathfrak{H} \odot \overline{\mathfrak{H}}$ satisfies the condition, $\alpha'' = \alpha$.

Proof. Let Φ be the symmetric gauge function generated by α (Lemma 5.28). Then, by Lemma 5.37, α' generates the associate of Φ, and again α'' generates the associate of the associate of Φ, which coincides with Φ by Lemma 5.27. Thus, α'' must coincide with α since they generate the same Φ. This concludes the proof.

8. The space $(\mathfrak{H} \otimes_{\alpha} \overline{\mathfrak{H}})^*$.

LEMMA 5.38. Let α be a crossnorm on $\mathfrak{H} \odot \overline{\mathfrak{H}}$. An operator A on $\mathfrak{H}$ is of finite α-norm (Definition 3.2) if and only if,

$$\sup_{0 \neq X \in \mathfrak{H} \odot \overline{\mathfrak{H}}} \frac{|t(XA)|}{\alpha(X)} < +\infty .$$

Moreover, the last sup always represents $\| A \|_{\alpha}$.

Proof. By Definition 3.2, an operator on $\mathfrak{H}$, that is, from $\mathfrak{H}$ into $\overline{\overline{\mathfrak{H}}}$ is of finite α-norm if and only if, there exists a finite constant c such that

$$\left| \sum_{i=1}^{m} (A\varphi_i, \psi_i) \right| \leqslant c\, \alpha\left(\sum_{i=1}^{m} \varphi_i \otimes \overline{\psi}_i \right)$$

for every operator $\sum_{i=1}^{m} \varphi_i \otimes \overline{\psi}_i$ in $\mathfrak{H} \odot \overline{\mathfrak{H}}$; the least of such constants is denoted by $\| A \|_{\alpha}$. Now if $X = \sum_{i=1}^{m} \varphi_i \otimes \overline{\psi}_i$, then $AX = \sum_{i=1}^{m} A\varphi_i \otimes \overline{\psi}_i$. By Lemma 5.12 and 5.15, $t(AX) = \sum_{i=1}^{m} (A\varphi_i, \psi_i)$. Hence also $t(XA) = \sum_{i=1}^{m} (A\varphi_i, \psi_i)$ by Lemma 5.13 (iv). Thus, above sup coincides with $\| A \|_{\alpha}$ of Definition 3.2. This concludes the proof.

The theorem which follows is a particular case of Theorem 3.1. Due to its frequent applications we restate it here in the language of unitary spaces.

THEOREM 5.6. Let $\mathfrak{F} \in (\mathfrak{H} \otimes_\alpha \overline{\mathfrak{H}})^*$. Then, there exists an operator A on $\mathfrak{H}$ of finite α-norm, satisfying the following conditions:

(i) $\mathfrak{F}(X) = t(XA)$ for $X \in \mathfrak{H} \odot \overline{\mathfrak{H}}$.

(ii) $\|A\|_\alpha = |||\mathfrak{F}|||$.

Conversely, whenever for an operator A on $\mathfrak{H}$ of finite α-norm $\mathfrak{F}$ is defined by means of (i), we have $\mathfrak{F} \in (\mathfrak{H} \otimes_\alpha \overline{\mathfrak{H}})^*$ and (ii) also holds.

Proof. Let $\mathfrak{F}$ denote an additive and bounded functional on $\mathfrak{H} \otimes_\alpha \overline{\mathfrak{H}}$. Clearly,

$$\mathfrak{F}(\varphi \otimes \overline{a\psi}) = \mathfrak{F}(\bar{a}(\varphi \otimes \overline{\psi})) = \bar{a}\,\mathfrak{F}(\varphi \otimes \overline{\psi})$$

and

$$\mathfrak{F}(\varphi \otimes \overline{\psi_1 + \psi_2}) = \mathfrak{F}(\varphi \otimes \overline{\psi_1}) + \mathfrak{F}(\varphi \otimes \overline{\psi_2}) .$$

Thus, holding φ fixed and varying ψ we see that $\mathfrak{F}(\varphi \otimes \overline{\psi})$ is the complex conjugate of an additive and bounded functional on $\mathfrak{H}$. Hence by a well-known Lemma of F. Riesz there exists a unique element, term it φ', such that $\mathfrak{F}(\varphi \otimes \overline{\psi}) = (\varphi', \psi)$. Define A by $A\varphi = \varphi'$. Thus,

$$\mathfrak{F}(\varphi \otimes \overline{\psi}) = (A\varphi, \psi) \qquad \text{for all } \varphi, \psi \text{ in } \mathfrak{H} .$$

Now for $X = \sum_{i=1}^{m} \varphi_i \otimes \overline{\psi_i}$ we have $\sum_{i=1}^{m} (A\varphi_i, \psi_i) = t(XA)$ by an argument in the proof of Lemma 5.38. Consequently,

$$\mathfrak{F}(X) = \sum_{i=1}^{m} \mathfrak{F}(\varphi_i \otimes \overline{\psi_i}) = \sum_{i=1}^{m} (A\varphi_i, \psi_i) = t(XA) .$$

Thus (i) holds. A is clearly additive. It is also bounded since by Lemma 3.2 $|||A||| \leq \|A\|_\alpha$ and

$$\|A\|_\alpha = \sup_{0 \neq X \in \mathfrak{H} \odot \overline{\mathfrak{H}}} \frac{|t(XA)|}{\alpha(X)} = \sup_{0 \neq X \in \mathfrak{H} \odot \overline{\mathfrak{H}}} \frac{|\mathfrak{F}(X)|}{\alpha(X)} = |||\mathfrak{F}||| .$$

Thus (ii) holds.

Conversely. Let $\|A\|_\alpha < +\infty$. Define $\mathfrak{F}$ by means of (i). Then, $\mathfrak{F}$ is additive and $\|A\|_\alpha$ represents its bound on $\mathfrak{K}\odot_\alpha\overline{\mathfrak{K}}$. Clearly, $\mathfrak{F}$ can be extended uniquely without changing its bound to an additive functional on $\mathfrak{K}\otimes_\alpha\overline{\mathfrak{K}}$. This concludes the proof.

THEOREM 5.7. For any crossnorm α, the Banach space $(\mathfrak{K}\otimes_\alpha\overline{\mathfrak{K}})^*$ may be interpreted as the space of all operators of finite α-norm on $\mathfrak{K}$, where

$$\|A\|_\alpha = \sup_{0\neq X\in\mathfrak{K}\odot\overline{\mathfrak{K}}} \frac{|t(XA)|}{\alpha(X)}$$

represents the norm of A.

Proof. The proof is a consequence of Theorem 5.6 and Lemma 5.38.

9. The Schmidt-class of operators as the cross-space $\mathfrak{K}\otimes_\sigma\overline{\mathfrak{K}} = (\mathfrak{K}\otimes_\sigma\overline{\mathfrak{K}})^*$.

We consider (sc). By Remark 5.1, (sc) is a linear space on which there is defined an inner product (X, Y) (Definition 5.2), with $\sigma(X) = (X, X)^{\frac{1}{2}}$ as the norm that goes with it. In particular, $\sigma(X)$ is also a norm for all operators X of finite rank, that is on $\mathfrak{K}\odot\overline{\mathfrak{K}}$. Moreover, σ has also the cross-property, since by Remark 5.2,

$$(\varphi\otimes\overline{\psi}, \varphi\otimes\overline{\psi})^{\frac{1}{2}} = ((\varphi, \varphi)(\psi, \psi))^{\frac{1}{2}} = \|\varphi\|\,\|\psi\| .$$

Thus the normed linear space $\mathfrak{K}\odot_\sigma\overline{\mathfrak{K}}$ is included in (sc).

The characterization of the crossnorm σ which follows attributes it a special significance.

DEFINITION 5.9. A crossnorm α is termed "self-associate" if $\alpha'(X) = \alpha(X)$ identically (for all operators X of finite rank).

LEMMA 5.39. σ is the unique self-associate crossnorm.

Proof. We shall prove first that $\sigma = \sigma'$. Let A be a fixed and X a variable operator of finite rank. Schwarz's inequality (Remark 5.1) gives,

$$|(A, X)| \leqslant \sigma(A)\, \sigma(X) \qquad \text{for all } X \text{ in } \mathfrak{H} \odot \overline{\mathfrak{H}}\,.$$

Thus, Definition 5.8 gives $\sigma'(A) \leqslant \sigma(A)$. On the other hand obviously $(\sigma(A))^2 = (A, A) \leqslant \sigma'(A)\, \sigma(A)$, hence $\sigma(A) \leqslant \sigma'(A)$. Thus, $\sigma(A) = \sigma'(A)$, that is, σ is a self-associate crossnorm in the sense of Definition 5.9.

To prove the uniqueness we first remark that the definition of the associate for a given crossnorm implies

$$(\sigma(A))^2 = (A, A) \leqslant \alpha(A)\, \alpha'(A)$$

for any operator A of finite rank. Assuming thus, $\alpha' = \alpha$ identically, we get $\sigma(A) \leqslant \alpha(A)$. Applying Lemma 2.2, we get $\alpha'(A) \leqslant \sigma'(A)$ and therefore also $\alpha(A) \leqslant \sigma(A)$ since by assumption $\alpha = \alpha'$ and $\sigma = \sigma'$. Thus, $\alpha(A) = \sigma(A)$ for all operators of finite rank. This concludes the proof.

THEOREM 5.8. $(\mathfrak{H} \otimes_\sigma \overline{\mathfrak{H}})^*$ may be considered as the Schmidt-class of operators on $\mathfrak{H}$.

Proof. We have pointed out before (Remark 5.1) that (sc) is a normed linear space. By Theorem 5.7 it is sufficient to show that an operator is in (sc) if and only if, it is of finite σ-norm, that is, if and only if,

$$\| A \|_\sigma = \sup_{0 \neq X \in \mathcal{K} \odot \bar{\mathcal{K}}} \frac{|t(XA)|}{\sigma(X)} < +\infty$$

and also

$$\| A \|_\sigma = \sigma(A) .$$

Let A be an operator $\neq 0$ for which $\| A \|_\sigma < +\infty$, and (φ_i) stand for a cnos. Since $A \neq 0$, for sufficiently large n we have, $\sum_{i=1}^{n} \| A\varphi_i \|^2 > 0$. Consider $\sum_{i=1}^{n} \varphi_i \otimes \overline{A\varphi_i}$. We have,

$$\sigma\left(\sum_{i=1}^{n} \varphi_i \otimes \overline{A\varphi_i} \right) = \left(\sum_{i=1}^{n} \| A\varphi_i \|^2 \right)^{\frac{1}{2}}$$

and

$$\sum_{i=1}^{n} \| A\varphi_i \|^2 = \sum_{i=1}^{n} (A\varphi_i , A\varphi_i) \leq$$
$$\| A \|_\sigma \, \sigma\left(\sum_{i=1}^{n} \varphi_i \otimes \overline{A\varphi_i} \right) = \| A \|_\sigma \left(\sum_{i=1}^{n} \| A\varphi_i \|^2 \right)^{\frac{1}{2}} .$$

Consequently,

$$\left(\sum_{i=1}^{n} \| A\varphi_i \|^2 \right)^{\frac{1}{2}} \leq \| A \|_\sigma \quad \text{for} \quad n = 1,2,\ldots.$$

that is, $A \in$ (sc) and $\sigma(A) = \left(\sum_i \| A\varphi_i \|^2 \right)^{\frac{1}{2}} \leq \| A \|_\sigma$.

On the other hand, let $A \in$ (sc) and X be any operator of finite rank. By Schwarz's inequality:

$$| t(XA) | = | (A , X^*) | \leq \sigma(A) \, \sigma(X^*) = \sigma(A) \, \sigma(X) .$$

Thus, Lemma 5.38 gives $\| A \|_\sigma \leq \sigma(A)$. Therefore, $\| A \|_\sigma < +\infty$ implies $A \in$ (sc) and $\| A \|_\sigma = \sigma(A)$.

Conversely, if $A \in$ (sc), $\sigma(A)$ is defined and finite. Then, the second half of the above proof, that is, Schwarz's inequality implies $\| A \|_\sigma \leq \sigma(A)$, that is, A is of finite σ-norm, and thus, by the first part of our proof $\| A \|_\sigma = \sigma(A)$. This concludes the proof.

THEOREM 5.9. We have $(\mathfrak{H} \otimes_{\sigma} \overline{\mathfrak{H}})^{*} = \mathfrak{H} \otimes_{\sigma} \overline{\mathfrak{H}}$; it represents the n^2-dimensional Euclidean space or a Hilbert space according to whether $\mathfrak{H}$ is an n-dimensional Euclidean space or a Hilbert space.

Proof. It is clearly sufficient to consider the case when $\mathfrak{H}$ is a Hilbert space.

Let $A \in (sc)$. It is "generally" known that an operator in (sc) is completely continuous. By Lemma 5.5, $A = \sum_i a_i \varphi_i \otimes \overline{\psi_i}$ where $a_i > 0$, $\lim a_i = 0$, and both (φ_i) and (ψ_i) are nos. Clearly (Theorem 5.8),

$$\sigma(A) = (\sum_i a_i^2)^{\frac{1}{2}} = \| A \|_{\sigma} < +\infty$$

Thus, for a given $\varepsilon > 0$, we can find an N for which

$$\| A - \sum_{i=1}^{N} a_i \varphi_i \otimes \overline{\psi_i} \|_{\sigma} = (\sum_{i>N} a_i^2)^{\frac{1}{2}} < \varepsilon .$$

This proves that every operator in (sc) hence by Theorem 5.8 of finite σ-norm can be approximated in that norm by operators of finite rank. Thus, $(\mathfrak{H} \otimes_{\sigma} \overline{\mathfrak{H}})^{*} = \mathfrak{H} \otimes_{\sigma'} \overline{\mathfrak{H}} = \mathfrak{H} \otimes_{\sigma} \overline{\mathfrak{H}}$ (Theorem 3.6).

Now the separability of $\mathfrak{H}$ implies the separability of $\mathfrak{H} \otimes_{\sigma} \overline{\mathfrak{H}}$. To see this we recall Lemma 2.4, which states that for any crossnorm α the normed linear space $\mathfrak{H} \odot_{\alpha} \overline{\mathfrak{H}}$ and hence $\mathfrak{H} \otimes_{\alpha} \overline{\mathfrak{H}}$ are separable. Moreover, for a cnos (φ_i) in $\mathfrak{H}$, the operators $\varphi_i \otimes \overline{\varphi_j}$; $i, j = 1, 2, \ldots$ form a cnos in $\mathfrak{H} \otimes_{\sigma} \overline{\mathfrak{H}}$. This concludes the proof.

It is not without interest to add (as follows from a construction in Appendix II) that σ may be obtained from <u>any</u> crossnorm by means of a sequence of operations, of taking the associate, using the arithmetic mean of two crossnorms and taking the limit of a monotonic sequence of crossnorms.

10. The trace-class as the cross-space $(\mathfrak{H}\otimes_\lambda\overline{\mathfrak{H}})^* = \mathfrak{H}\otimes_\gamma\overline{\mathfrak{H}}$.

We recall that for an operator A of finite rank, the symbols $\lambda(A)$ and $\gamma(A)$ stand for its bound and value assumed for the greatest crossnorm, respectively. The symbol m(A) stands for t(abs(A)) (Definition 5.4).

LEMMA 5.40. For any operator A of finite rank $m(A) = \gamma(A)$. The crossnorms γ and λ are associate with each other, that is, $\gamma' = \lambda$ and $\lambda' = \gamma$.

Proof. By Remark 5.5, m(A) is a crossnorm on $\mathfrak{H}\odot\overline{\mathfrak{H}}$. Hence, $m(A) \leq \gamma(A)$ for all operators A of finite rank. To prove the converse inequality, we write A (by Lemma 5.5) in the canonical form

$$A = \Sigma_{i=1}^{m} a_i \varphi_i \otimes \overline{\psi}_i$$

where the a_i's are > 0 and the φ_i's and ψ_i's form nos. We have,

$$A^*A = \Sigma_{i=1}^{m} a_i^2 \psi_i \otimes \overline{\psi}_i$$

Extending $\psi_1, \ldots\ldots, \psi_m$ to a cnos (ψ_i) we see that

$\psi_1, \psi_2, \psi_3, \ldots\ldots, \psi_m, \psi_{m+1}, \psi_{m+2}, \ldots\ldots$ are proper vectors of A^*A corresponding to the proper values $a_1^2, a_2^2, a_3^2, \ldots\ldots, a_m^2, 0, 0, \ldots.$. Therefore, they are also proper vectors of abs(A) corresponding to the proper values $a_1, a_2, a_3, \ldots\ldots, a_m, 0, 0, \ldots.$. Consequently,

$$m(A) = \Sigma_i (\text{abs}(A)\psi_i, \psi_i) = \Sigma_i (a_i\psi_i, \psi_i) = \Sigma_{i=1}^{m} a_i .$$

The sum on the extreme right is clearly $\geq \gamma(A)$ since by definition,

$$\gamma(A) = \gamma(\Sigma_{i=1}^{m} a_i \varphi_i \otimes \overline{\psi}_i) \leq \Sigma_{i=1}^{m} \| a_i\varphi_i \| \|\psi_i\| = \Sigma_{i=1}^{m} a_i .$$

Thus, $\gamma(A) \leq m(A)$ and therefore,

$$\gamma(A) = m(A) \qquad \text{for all } A \text{ in } \mathfrak{H}\odot\overline{\mathfrak{H}} .$$

It remains to prove the last statement of our Lemma. By Lemma 5.3, we have $\lambda(\sum_{i=1}^{m} \varphi_i \otimes \bar{\varphi}_i) = 1$. The definition of the associate λ' gives,

$$\lambda'(A) = \lambda'(\sum_{i=1}^{m} a_i \varphi_i \otimes \bar{\varphi}_i) =$$

$$\lambda'(\sum_{i=1}^{m} a_i \varphi_i \otimes \bar{\varphi}_i)\ \lambda(\sum_{i=1}^{m} \varphi_i \otimes \bar{\varphi}_i) \geq$$

$$(\sum_{i=1}^{m} a_i \varphi_i \otimes \bar{\varphi}_i,\ \sum_{i=1}^{m} \varphi_i \otimes \bar{\varphi}_i) = \sum_{i=1}^{m} a_i = \gamma(A) ;$$

that is, $\lambda'(A) \geq \gamma(A)$. Since λ' is a crossnorm, we have also $\lambda' \leq \gamma$. Thus, $\lambda'(A) = \gamma(A)$.

That $\gamma' = \lambda$ holds even for general Banach spaces has been proven in Theorem 2.5. This concludes the proof.

We are about to consider the Banach space of all completely continuous operators on $\mathfrak{H}$, and characterize its first and second conjugate space.

THEOREM 5.10. The linear space of all completely continuous operators, with the bound of an operator as its norm, furnishes the Banach space $\mathfrak{H} \otimes_{\lambda} \bar{\mathfrak{H}}$.

Proof. Clearly, every operator X of finite rank is completely continuous; $\lambda(X)$ represents its bound. Hence, $\mathfrak{H} \odot_{\lambda} \bar{\mathfrak{H}}$ represent the normed linear space of operators of finite rank, where the bound of an operator stands for its norm. To complete our proof it is sufficient to show that an operator on $\mathfrak{H}$ is completely continuous if and only if, it can be approximated in bound by a sequence of operators of finite rank.

So let A represent a completely continuous operator. By Lemma 5.5, $A = \sum_i a_i \varphi_i \otimes \bar{\psi}_i$ where $a_i > 0$, $\lim a_i = 0$, and both the φ_i's as well as the ψ_i's form nos. By Lemma 5.3, the bound of $A - \sum_{i=1}^{m} a_i \varphi_i \otimes \bar{\psi}_i$

is given by

$$\sup_{i > n} a_i .$$

Since $\lim a_i = 0$, the last number obviously approaches 0 as $n \to \infty$.

Conversely, the limit of a sequence of completely continuous (hence also of finite rank) operators convergent in bound is also a completely continuous operator by [1, p. 96]. This concludes the proof.

The trace-class (tc) has been defined before (Definition 5.3). By Remark 5.5, (tc) forms a linear space on which $m(A)$ is a crossnorm.

THEOREM 5.11. The Banach space $(\mathfrak{H} \otimes_\lambda \bar{\mathfrak{H}})^*$ that is, the conjugate space of the space of all completely continuous operators, may be interpreted as the trace-class, that is, the space of all operators A on $\mathfrak{H}$ for which $\sum_i ((A^*A)^{\frac{1}{2}} \varphi_i , \varphi_i) < +\infty$ for a cnos (φ_i), and where the last infinite sum which is independent on the chosen cnos represents the norm of A.

Proof. By Theorem 5.7, it is sufficient to prove that the trace-class represents the space of all those operators A for which

$$\text{(i)} \qquad \| A \|_\lambda = \sup_{0 \neq X \in \mathfrak{H} \odot \bar{\mathfrak{H}}} \frac{| t(XA) |}{\lambda(X)} < +\infty$$

and

$$\text{(ii)} \qquad \| A \|_\lambda = m(A) .$$

We assume first that for an operator $A \neq 0$ the number $\| A \|_\lambda$ is finite. By Lemma 5.1 (ii), there exists a partial isometric operator W whose initial set is the closed linear manifold determined by the range of $\mathrm{abs}(A)$ for which $\mathrm{abs}(A) = W^*A$. Let $\psi_1, \psi_2, \psi_3, \ldots$ denote a cnos in the closure of the range of $\mathrm{abs}(A)$. We extend it to $\psi_1, \psi_2, \psi_3, \ldots, \omega_1, \omega_2, \omega_3,$

a cnos in $\mathfrak{K}$, which we shall denote by (φ_i). We may also assume $\varphi_1 = \psi_1$. Clearly, the $W\psi_i$'s form an orthonormal set, while $W\omega_j = 0$. By Lemma 5.3, for every natural p we have, $\lambda(\Sigma_{i=1}^{p} \varphi_i \otimes \overline{W\varphi_i}) = 1$.

Therefore,

$$\Sigma_{i=1}^{p}(\mathrm{abs}(A)\varphi_i, \varphi_i) = \Sigma_{i=1}^{p}(W^*A\varphi_i, \varphi_i) =$$

$$\Sigma_{i=1}^{p}(A\varphi_i, W\varphi_i) \leqslant \|A\|_\lambda \; \lambda(\Sigma_{i=1}^{p} \varphi_i \otimes \overline{W\varphi_i}) = \|A\|_\lambda .$$

The last inequality holds for $p = 1, 2, \ldots$. Thus, abs(A) and therefore also $A \in (tc)$ and

$$m(A) = t(\mathrm{abs}(A)) = \Sigma_i (\mathrm{abs}(A)\varphi_i, \varphi_i) \leqq \|A\|_\lambda .$$

On the other hand for any X of finite rank, Lemma 5.14 (vi) and (v) gives,

$$|t(XA)| \leqslant m(XA) \leqslant |||X||| \, m(A) = \lambda(X) \, m(A) .$$

This implies, $\|A\|_\lambda \leqslant m(A)$. Thus, $\|A\|_\lambda < +\infty$ implies $A \in (tc)$ and $\|A\|_\lambda = m(A)$.

Conversely, $A \in (tc)$ implies by Lemma 5.11, $\mathrm{abs}(A) \in (tc)$ and thus m(A) is defined and finite. The second half of the above proof gives, $\|A\|_\lambda \leqslant m(A) < +\infty$, and consequently $\|A\|_\lambda = m(A)$ by the first part. This concludes the proof.

COROLLARY 5.1. The trace-class of operators A is complete relative to the m(A) norm.

Proof. This is a consequence of Theorem 5.11.

LEMMA 5.41. For $A \in (tc)$, abs(A) possesses a pure point spectrum. If $a_1, a_2, \ldots$ denote the point proper values (with multiplicities) of abs(A), then $m(A) = \Sigma_i a_i < +\infty$.

Proof. Let $A \in (tc)$. By Lemma 5.11 this is equivalent to $(\mathrm{abs}(A))^{\frac{1}{2}} \in (sc)$. Now it is well-known that every definite Hermitean operator in (sc) possesses a pure point spectrum, i.e., there exists a cnos consisting of proper vectors of this operator. Let (φ_i) be such a cnos for $(\mathrm{abs}(A))^{\frac{1}{2}}$. Each φ_i is a proper vector for $(\mathrm{abs}(A))^{\frac{1}{2}}$ hence also for $\mathrm{abs}(A)$. Let a_i be the proper value of $\mathrm{abs}(A)$ corresponding to the proper vector φ_i. Clearly, we have

$$\Sigma_i a_i = \Sigma_i (\mathrm{abs}(A)\varphi_i, \varphi_i) = m(A) < +\infty .$$

This concludes the proof.

LEMMA 5.42. The set of all operators of finite rank X is dense in (tc) relative to the $m(X)$ norm.

Proof. Let $A \in (tc)$. We use Lemma 5.41 and the notation in its proof. For an $\varepsilon > 0$ we can find a finite p such that $\Sigma_{i>p} a_i \leq \varepsilon$. We form two operators B and C, which we obtain from $\mathrm{abs}(A)$ by replacing its proper values a_i by 0 for $i \leq p$ or for $i > p$, by leaving its proper values a_i unchanged for $i > p$ or for $i \leq p$, and also by leaving the corresponding proper vectors φ_i unchanged in all cases. Then,

$$\mathrm{abs}(A) = B + C .$$

B has finite rank and both B and C are Hermitean and definite along with $\mathrm{abs}(A)$. The above properties of C imply

$$\mathrm{abs}(C) = (C^*C)^{\frac{1}{2}} = (C^2)^{\frac{1}{2}} = C$$

and therefore,

$$\Sigma_i (\mathrm{abs}(C)\varphi_i , \varphi_i) = \Sigma_i (C\varphi_i , \varphi_i) = \Sigma_{i>p} a_i \leq \varepsilon .$$

Thus, $C \in (tc)$ and $m(C) = t(C) \leq \varepsilon$.

By Lemma 5.1 (i), $A = W\mathrm{abs}(A)$, $|||W||| = 1$. Therefore, $A = WB + WC$. It is clear that WB is of finite rank along with B , and

$$m(A - WB) = m(WC) \leq |||W||| \, m(C) \leq \varepsilon .$$

This concludes the proof.

THEOREM 5.12. The trace-class may be also interpreted as $\mathfrak{K} \otimes_\gamma \overline{\mathfrak{K}}$.

Proof. By Corollary 5.1, the trace-class of operators is complete relative to the norm $m(A)$. In it by Lemma 5.42, the operators of finite rank form a dense linear subspace. This subspace coincides with $\mathfrak{K} \odot_\gamma \overline{\mathfrak{K}}$, since by Lemma 5.40, we have $\gamma(A) = m(A)$ for all operators A of finite rank. Consequently, the trace-class coincides with the smallest Banach space in which $\mathfrak{K} \odot_\gamma \overline{\mathfrak{K}}$ can be imbedded, that is, with $\mathfrak{K} \otimes_\gamma \overline{\mathfrak{K}}$
This concludes the proof.

THEOREM 5.13. $(\mathfrak{K} \otimes_\lambda \overline{\mathfrak{K}})^* = \mathfrak{K} \otimes_\gamma \overline{\mathfrak{K}}$.

Proof. This is a consequence of Theorems 5.11 and 5.12.

THEOREM 5.14. $(\mathfrak{K} \otimes_\gamma \overline{\mathfrak{K}})^*$ may be interpreted as the Banach space of all operators on $\mathfrak{K}$ where the norm of an operator is given by its bound.

Proof. This is a special case of Theorem 3.2.

THEOREM 5.15. The linear space of all completely continuous operators on $\mathfrak{H}$ where the bound of an operator is considered as its norm, furnishes a Banach space $\mathfrak{C}$. Its first conjugate space $\mathfrak{C}^*$ may be interpreted as the trace-class. The trace-class is the Banach space of all operators A on $\mathfrak{H}$, for which $\sum_i ((A^*A)^{\frac{1}{2}}\varphi_i, \varphi_i) < +\infty$ for a complete orthonormal set (φ_i); the last sum which is independent on the chosen complete orthonormal set represents the norm of A in $\mathfrak{C}^*$. The second conjugate space $\mathfrak{C}^{**}$ of $\mathfrak{C}$ may be interpreted as the Banach space of all operators on $\mathfrak{H}$ where again the bound of an operator represents its norm.

Moreover, $\mathfrak{C}$ may be characterized as $\mathfrak{H}\otimes_\lambda\bar{\mathfrak{H}}$, while the space $\mathfrak{C}^*$ may be interpreted as $(\mathfrak{H}\otimes_\lambda\bar{\mathfrak{H}})^* = \mathfrak{H}\otimes_\gamma\bar{\mathfrak{H}}$, and finally $\mathfrak{C}^{**}$ as $(\mathfrak{H}\otimes_\lambda\bar{\mathfrak{H}})^{**} = (\mathfrak{H}\otimes_\gamma\bar{\mathfrak{H}})^*$.

Proof. The proof is a consequence of Theorems 5.10, 5.11, 5.12 and 5.14.

COROLLARY 5.2. $\mathfrak{H}\otimes_\lambda\bar{\mathfrak{H}}$ is a proper subspace of $(\mathfrak{H}\otimes_\gamma\bar{\mathfrak{H}})^*$.

Proof. By Theorem 2.5, $\gamma' = \lambda$. Hence, Theorems 5.10 and 5.14 furnish the desired proof.

We conclude this section with a few words about the inclusion $\mathfrak{H}\otimes_{\alpha'}\bar{\mathfrak{H}} \subset (\mathfrak{H}\otimes_\alpha\bar{\mathfrak{H}})^*$ for any "limited" crossnorm α.

DEFINITION 5.10. Let α denote a crossnorm on $\mathfrak{H}\odot\bar{\mathfrak{H}}$ and p a natural number. For an operator A of finite rank we define

$$\alpha_p(A) = \sup \frac{|(A, X)|}{\alpha(X)}$$

where the sup is taken over the set of all operators X of rank $\leqslant$ p.

LEMMA 5.43.

(i) All α_p are reflexive crossnorms.

(ii) $\alpha_1 \leq \alpha_2 \leq \alpha_3 \leq \ldots$.

(iii) $\alpha_1 = \lambda$.

(iv) $\lim_{p \to \infty} \alpha_p = \alpha'$.

(v) $\alpha \geq \beta$ implies $\alpha_p \leq \beta_p$.

Proof. The proof of (i)--(iv) is presented in Theorem 4 of Appendix I, while (v) is an immediate consequence of Definition 5.10.

LEMMA 5.44. $\lambda_p \leq p\lambda$ for $p = 1,2,\ldots\ldots$

Proof. Let A be a fixed operator of finite rank. We present the variable operator X of rank $r \leq p$ in the canonical form (Lemma 5.5),

$$X = \Sigma_{i=1}^{r} x_i \varphi_i \otimes \overline{\psi_i} = \Sigma_{i=1}^{r} X_i$$

where $X_i = x_i \varphi_i \otimes \overline{\psi_i}$. The x_i's are > 0 , and both the (φ_i) as well as the (ψ_i) form nos. By Lemma 5.3,

$$|||X_i||| = x_i \leq \max_{1 \leq i \leq r} x_i = |||X||| .$$

Since λ and γ are associate with each other (Lemma 5.40) we have,

$$|(A , X_i)| \leq \lambda(A)\ \gamma(X_i) =$$
$$|||A|||\ |||X_i||| \leq |||A|||\ |||X|||$$

and therefore,

$$|(A , X)| = |(A , \Sigma_{i=1}^{r} X_i)| \leq \Sigma_{i=1}^{r} |(A , X_i)| \leq$$
$$r\,|||A|||\ |||X||| \leq p\ \lambda(A)\ \lambda(X) .$$

Thus, Definition 5.10 gives, $\lambda_p(A) \leq p\,\lambda(A)$. This concludes the proof.

LEMMA 5.45. For any crossnorm $\alpha \geqslant \lambda$, we have $\alpha_p \leqslant p\lambda$.

Proof. By assumption $\alpha \geqslant \lambda$. By Lemma 5.43 (v), $\alpha_p \leqslant \lambda_p$. Thus, $\alpha_p \leqslant p\lambda$ by Lemma 5.44.

THEOREM 5.16. For a "limited" crossnorm α, the following relationships hold:

(i) $(\mathfrak{H} \otimes_\alpha \overline{\mathfrak{H}})^* = \mathfrak{H} \otimes_{\alpha'} \overline{\mathfrak{H}}$.

(ii) $\mathfrak{H} \otimes_\alpha \overline{\mathfrak{H}}$ is a proper subspace of $(\mathfrak{H} \otimes_{\alpha'} \overline{\mathfrak{H}})^*$.

Proof. By Lemmas 5.43 (iii) and 5.45, for a certain natural p we have $\lambda \leqslant \alpha \leqslant p\lambda$. This proves that the spaces $\mathfrak{H} \otimes_\lambda \overline{\mathfrak{H}}$ and $\mathfrak{H} \otimes_\alpha \overline{\mathfrak{H}}$ are topologically equivalent. Thus, (i) is a consequence of Theorem 5.13, while (ii) follows from the fact that $\mathfrak{H} \otimes_\lambda \overline{\mathfrak{H}}$ is a proper subspace of $(\mathfrak{H} \otimes_\lambda \overline{\mathfrak{H}})^* = (\mathfrak{H} \otimes_\gamma \overline{\mathfrak{H}})^*$ (Theorem 5.15). This concludes the proof.

COROLLARY 5.3. For a limited crossnorm α the spaces $\mathfrak{H} \otimes_\alpha \overline{\mathfrak{H}}$ and $\mathfrak{H} \otimes_{\alpha'} \overline{\mathfrak{H}}$ are non-reflexive.

Proof. This follows from Theorem 2 of Appendix I.

The following illustration is not without interest: For λ we construct the corresponding sequence $\lambda_1, \lambda_2, \ldots$ of limited crossnorms. Put $\lim_{p \to \infty} \lambda_p = \lambda_\infty$. By Lemma 5.43, $\lambda_\infty = \lambda' = \gamma$. Furthermore, $\lim_{p \to \infty} (\lambda_p)' = (\lambda_\infty)' = \lambda$. By Theorem 5.16, for every natural p the conjugate and associate space for the cross-space $\mathfrak{H} \otimes_{\lambda_p} \overline{\mathfrak{H}}$ coincide, while for $p = \infty$ the associate space is a proper subspace of the conjugate space (Corollary 5.2).

Moreover, for every natural p , $\mathfrak{H} \otimes_{\lambda_p} \overline{\mathfrak{H}}$ is a proper subspace of $(\mathfrak{H} \otimes_{(\lambda_p)'} \overline{\mathfrak{H}})^*$, while for $p = \infty$ they coincide (Theorem 5.13).

11. The structure of ideals $(\mathfrak{H} \otimes_\alpha \overline{\mathfrak{H}})^*$.

"Algebraic ideals" of operators on $\mathfrak{H}$ have been considered in the literature. A linear set $\mathfrak{B}$ of operators A on $\mathfrak{H}$ is termed an algebraic ideal if

(i) $A \in \mathfrak{B}$ implies $YAX \in \mathfrak{B}$ for any pair of operators X , Y.

Concerning these algebraic ideals of interest is the following theorem whose proof may be found in [2a, p. 841] :

An algebraic ideal which does not include all operators consists solely of completely continuous operators.

The notion of an "ideal" of operators on $\mathfrak{H}$ (and also from one Banach space into another) has been presented here in Definition 4.1. A Banach space $\mathfrak{B}$ of operators is an ideal if in addition to condition (i) above, the following condition holds:

(ii) $\|YAX\| \leq |||Y|||\ \|A\|\ |||X|||$

where $\|A\|$ represents the norm of A in $\mathfrak{B}$.

Clearly, an ideal is also an algebraic ideal. Thus, whenever an ideal does not include all operators, all its elements must be completely continuous. An algebraic ideal however, may not be an ideal as follows from the following argument:

Let $\mathfrak{H}_2$ denote the two dimensional Euclidean space and α a non-unitarily invariant (non-uniform) crossnorm on $\mathfrak{H}_2 \odot \overline{\mathfrak{H}}_2$. Such crossnorms will be constructed in the last section of this Chapter. Clearly, $(\mathfrak{H}_2 \otimes_\alpha \overline{\mathfrak{H}}_2)^*$ that is, the linear space of all operators on $\mathfrak{H}_2$ is an "algebraic ideal". However, it is not an ideal, since by Theorem 5.3, $(\mathfrak{H}_2 \otimes_\alpha \overline{\mathfrak{H}}_2)^*$ is an ideal if and only if α is unitarily invariant.

DEFINITION 5.11. A unitarily invariant crossnorm α will be termed "significant" if every operator of finite α-norm is completely continuous.

It follows that a unitarily invariant crossnorm α is significant if and only if, there exists an operator which is not of finite α-norm. It is clear that for $\beta \leq \alpha$ the operators of finite β-norm are also of finite α-norm. Hence, whenever α is significant, β must be also significant. In particular, for instance, since the identity operator does not belong to the Schmidt-class, that is, is not of finite σ-norm, every unitarily invariant crossnorm $\alpha \leq \sigma$ is significant. Clearly, the greatest crossnorm γ is not significant, since every operator is of finite γ-norm (Lemma 3.5).

In the present section we shall discuss the structure of the conjugate spaces of cross-spaces generated by significant unitarily invariant crossnorms. In particular, we shall discuss the relationship to their associate spaces.

LEMMA 5.46. Let α denote a unitarily invariant crossnorm. An operator A is of finite α-norm if and only if, $\mathrm{abs}(A)$ is such. Moreover, $\| A \|_\alpha = \| \mathrm{abs}(A) \|_\alpha$.

Proof. By Theorem 5.3, the space of all operators of finite α-norm forms an ideal. By Lemma 5.1, $A = W\mathrm{abs}(A)$ and $\mathrm{abs}(A) = W^*A$.

We have,

$$\| \mathrm{abs}(A) \|_{\alpha} = \| W^{*}A \|_{\alpha} \leq |||W^{*}||| \; \|A\|_{\alpha} = \|A\|_{\alpha}$$

and

$$\|A\|_{\alpha} = \| W\mathrm{abs}(A) \|_{\alpha} \leq |||W||| \; \|\mathrm{abs}(A)\|_{\alpha} = \|\mathrm{abs}(A)\|_{\alpha} .$$

Therefore, $\|A\|_{\alpha} = \|\mathrm{abs}(A)\|_{\alpha}$. This concludes the proof.

THEOREM 5.17. Let α be a unitarily invariant crossnorm. A completely continuous operator A is of finite α-norm if and only if, its canonical representation $A = \sum_i a_i \varphi_i \otimes \overline{\psi}_i$ is either finite, or if infinite, then

(i) $$\lim_{n\to\infty} \alpha'(\textstyle\sum_{i=1}^{n} a_i \varphi_i \otimes \overline{\psi}_i) < +\infty .$$

Furthermore, in the last case

(ii) $$\lim_{n\to\infty} \alpha'(\textstyle\sum_{i=1}^{n} a_i \varphi_i \otimes \overline{\psi}_i) = \|A\|_{\alpha} .$$

Proof. Suppose that the completely continuous operator A is of finite α-norm. Put,

$$A = \textstyle\sum_i a_i \varphi_i \otimes \overline{\psi}_i \qquad \text{and} \qquad A_n = \sum_{i=1}^{n} a_i \varphi_i \otimes \overline{\psi}_i .$$

We remark that $B = \sum_i \varepsilon_i a_i \varphi_i \otimes \overline{\psi}_i$ with $\varepsilon_i = \pm 1$ is also of finite α-norm, since $\mathrm{abs}(A) = \mathrm{abs}(B)$ and therefore $\|A\|_{\alpha} = \|B\|_{\alpha}$ by Lemma 5.46. In particular, $2A_n - A$ is of finite α-norm and

$$\| 2A_n - A \|_{\alpha} = \| A \|_{\alpha}$$

Thus, for any natural n we have,

$$\|A_n\|_{\alpha} = \| \tfrac{1}{2}A + \tfrac{1}{2}(2A_n - A) \|_{\alpha} \leq$$
$$\tfrac{1}{2}\| A \|_{\alpha} + \tfrac{1}{2}\|2A_n - A \|_{\alpha} = \|A\|_{\alpha} .$$

By assumption α is unitarily invariant, hence α' is such by Theorem 5.2

and Lemma 5.37. Consequently, $\alpha'(A_n) = \|A_n\|_\alpha$ is a non-decreasing function of n (Lemma 5.16); by above inequality:

$$\lim_{n\to\infty} \alpha'(A_n) = \lim_{n\to\infty} \|A_n\|_\alpha \leq \|A\|_\alpha .$$

To prove the converse inequality, we remark that for any operator X of finite rank $t(XA)$ is defined and

$$t(XA_n) \longrightarrow t(XA) .$$

This follows from $\lim_i a_i = 0$ and

$$| t(XA_n) - t(XA)| = | t(X(A_n - A)| \leq$$
$$m(X) ||| A_n - A ||| = m(X) \sup_{i>n} a_i$$

Clearly, for every n we have,

$$| t(XA_n)| = |(A_n , X^*)| \leq$$
$$\alpha'(A_n) \alpha(X^*) \leq \lim_{n\to\infty} \alpha'(A_n) \alpha(X) .$$

Therefore, also in the limit

$$| t(XA)| \leq \lim_{n\to\infty} \alpha'(A_n) \alpha(X) .$$

Since the last inequality holds for all operators X of finite rank, Lemma 5.38 gives,

$$\| A \|_\alpha \leq \lim_{n\to\infty} \alpha'(A_n) .$$

This proves that for a completely continuous operator $\sum_i a_i \varphi_i \otimes \overline{\psi_i}$ of finite α-norm (i) and (ii) hold. Conversely, suppose that for a completely continuous operator $A = \sum_i a_i \varphi_i \otimes \overline{\psi_i}$ relation (i) holds. The last part of above proof furnishes $\| A \|_\alpha \leq \lim_{n\to\infty} \alpha'(A_n)$. Hence, A is of finite α-norm, and by the first part also (ii) holds. This concludes the proof.

THEOREM 5.17a. For a significant unitarily invariant crossnorm α (Definition 5.11), $(\mathfrak{K} \otimes_\alpha \overline{\mathfrak{K}})^*$ represents precisely the space of all those completely continuous operators whose canonical form is either finite, or if infinite then it satisfies (i). Furthermore, in the last case (ii) also holds.

THEOREM 5.18. We assume that (ψ_i) denotes any fixed -- throughout the discussion -- cnos in $\mathfrak{K}$. Let α denote a "significant" unitarily invariant crossnorm. Suppose that for every sequence of real numbers (a_i) with $\lim a_i = 0$, the condition

(i) $\qquad \lim \alpha'(\sum_{i=1}^{m} a_i \varphi_i \otimes \overline{\psi}_i)$

implies

(ii) $\qquad \lim\limits_{m,n \to \infty} \alpha'(\sum_{i=m}^{n} a_i \varphi_i \otimes \overline{\psi}_i) = 0 .$

Then, $(\mathfrak{K} \otimes_\alpha \overline{\mathfrak{K}})^* = \mathfrak{K} \otimes_{\alpha'} \overline{\mathfrak{K}}$.

Proof. Together with α, its associate α' is also a unitarily invariant crossnorm. Hence the value of $\alpha'(\sum_{i=1}^{m} a_i \varphi_i \otimes \overline{\psi}_i)$ does not depend on the normalized orthogonal sets (φ_i) and (ψ_i).

Let A be of finite α-norm and $\sum_i a_i \varphi_i \otimes \overline{\psi}_i$ its canonical representation of Lemma 5.5. Thus, $\lim a_i = 0$. By Theorem 5.17, the sequence $a_1, a_2, \ldots$ satisfies (i) above. This by our present assumption implies that (ii) holds. It follows that the sequence of operators

$A_n = \sum_{i=1}^{n} a_i \varphi_i \otimes \overline{\psi}_i \qquad n = 1,2,\ldots$ is fundamental in $\mathfrak{K} \otimes_{\alpha'} \overline{\mathfrak{K}}$,

and this determines a unique operator $\tilde{A}$, for which

$$\lim_{n \to \infty} \| \tilde{A} - A_n \|_\alpha = 0 .$$

Consequently,

$$\lim_{n \to \infty} |||\tilde{A} - A_n||| = 0 .$$

On the other hand,

$$\lim_{n \to \infty} ||| A - A_n ||| = \lim_{n \to \infty} (\sup_{i > n} a_i) = 0 .$$

Thus, $||| A - \tilde{A} ||| = 0$ or $A = \tilde{A}$.

Therefore, $(\mathfrak{H} \otimes_\alpha \bar{\mathfrak{H}})^* \subset \mathfrak{H} \otimes_{\alpha'} \bar{\mathfrak{H}}$. On the other hand since the converse inclusion always holds, we have $(\mathfrak{H} \otimes_\alpha \bar{\mathfrak{H}})^* = \mathfrak{H} \otimes_{\alpha'} \bar{\mathfrak{H}}$. This concludes the proof.

12. A crossnorm whose associate is not a crossnorm.

We conclude this Chapter with a proof that λ is not necessarily the least crossnorm. In fact, we shall prove that if $\mathfrak{H}_n$ denotes an n-dimensional Euclidean space, $n = 2,3,\ldots$ (but not 1!), there is no least crossnorm on $\mathfrak{H}_n \odot \bar{\mathfrak{H}}_n$, and we shall actually construct crossnorms which are not $\geqq \lambda$. These by Theorem 2.1 furnish examples of crossnorms (even on finite-dimensional spaces) whose associates are not crossnorms. By Theorem 5.4, none of these crossnorms is unitarily invariant, hence uniform.

Clearly, $\mathfrak{H}_n \odot \bar{\mathfrak{H}}_n$ represents the linear set of all operators on $\mathfrak{H}_n$. Let I stand for the identity operator.

DEFINITION 5.12. A non-negative function $\alpha(X)$ of operators X on $\mathfrak{H}_n$ is termed a quasi-norm if it satisfies conditions (ii) and (iii) of Definition 5.7. A quasi-norm is a quasi-crossnorm if it satisfies also (iv) of Definition 5.7.

LEMMA 5.47. Let X denote an operator of rank 1 on the two-dimensional Euclidean space $\mathfrak{R}_2$. Then, for any complex number a, we have $\gamma(X - aI) \geqslant \gamma(X)$.

Proof. By a suitable choice of the coordinate vectors φ_1, φ_2, in $\mathfrak{R}_2$ we can assure that the matrix of X has the form

$$\begin{pmatrix} 0 & d \\ 0 & c \end{pmatrix}$$

Since we may replace X by θX (θ a complex number with $|\theta| = 1$), we can make c real and $\geqslant 0$. Since we may replace φ_1 by $\tilde{\theta}\varphi_1$ ($\tilde{\theta}$ a complex number with $|\tilde{\theta}| = 1$), we may assume that d is real and $\geqslant 0$. Thus,

$$X = \begin{pmatrix} 0 & d \\ 0 & c \end{pmatrix} \qquad X - aI = \begin{pmatrix} -a & d \\ 0 & c-a \end{pmatrix}$$

where c, d, are real and $\geqslant 0$, while a is complex.

Let $t(X)$ and $|X|$ denote the trace and the determinant of X respectively. We have:

$$t(\,(X - aI)^*(X - aI)\,) = c^2 + d^2 + 2\,|a|^2 - 2c\,\mathcal{R}a\,.$$

$$|(X - aI)^*(X - aI)| = |a(a - c)|^2\,.$$

Let a_1, a_2, (real and $\geqslant 0$), represent the characteristic values of $\mathrm{abs}(X - aI)$. Then, by Lemma 5.40,

$$\gamma(X - aI) = a_1 + a_2\,.$$

Since a_1^2, a_2^2, are the characteristic values of $(X - aI)^*(X - aI)$ we have,

$$a_1^2 + a_2^2 = c^2 + d^2 + 2\,|a|^2 - 2c\,\mathcal{R}a\,,$$

and also $a_1^2\, a_2^2 = |a(a - c)|^2$, that is, $a_1\, a_2 = |a(a - c)|$.

Therefore,

$$\gamma(X - aI) = a_1 + a_2 = \sqrt{(a_1^2 + a_2^2) + 2a_1 a_2} =$$
$$\sqrt{c^2 + d^2 - 2c\mathcal{R}a + 2|a|^2 + 2|a(a-c)|} .$$

For $a = 0$, this becomes

$$\gamma(X) = \sqrt{c^2 + d^2} .$$

Thus, we wish to establish the following relation:

$$\sqrt{c^2 + d^2 - 2c\mathcal{R}a + 2|a|^2 + 2|a(a-c)|} \geqslant \sqrt{c^2 + d^2}$$

that is,

$$|a(a-c)| \geqslant c\mathcal{R}a - |a|^2 .$$

This however, is obvious, since

$$|a(a-c)| \geqslant |c|\,|a| - |a|^2 \geqslant c\mathcal{R}a - |a|^2 .$$

This concludes the proof.

LEMMA 5.48. Let X denote an operator of rank 1 defined on an n-dimensional Euclidean space $\mathfrak{R}_n$. Then for every complex number a, we have $\gamma(X - aI) \geqslant \gamma(X)$.

Proof. The case $n = 2$ has been proven in Lemma 5.47. We may assume $n > 2$. Since X is of rank 1, it is of the form $\varphi \otimes \overline{\psi}$.

Let $\mathfrak{R}'$ be a two-dimensional Euclidean space containing φ and ψ. Then, X is completely reduced by $\mathfrak{R}'$, that is, X (and of course I too) may be considered as an operator on $\mathfrak{R}'$ and as an operator on $\mathfrak{R}_n - \mathfrak{R}'$. In $\mathfrak{R}'$, the operator X still has rank 1, while in $\mathfrak{R}_n - \mathfrak{R}'$ the X is identically zero. Therefore,

$$\gamma(X - aI) = \gamma_{\mathfrak{R}'}(X - aI) + \gamma_{\mathfrak{R}_n - \mathfrak{R}'}(X - aI) \geqslant \gamma_{\mathfrak{R}'}(X - aI)$$

and

$$\gamma(X) = \gamma_{\mathfrak{K}'}(X) + \gamma_{\mathfrak{K}_m \ominus \mathfrak{K}'}(X) = \gamma_{\mathfrak{K}'}(X) .$$

Thus, it is sufficient to prove that

$$\gamma_{\mathfrak{K}'}(X - aI) \geqslant \gamma_{\mathfrak{K}'}(X)$$

In other words the whole problem may be considered entirely within $\mathfrak{K}'$. The last inequality holds by virtue of Lemma 5.47. This concludes the proof.

LEMMA 5.49. There exists a quasi-crossnorm ρ defined on $\mathfrak{K}_m \odot \overline{\mathfrak{K}}_m$ such that $\rho(I) = 0$.

Proof. We put

$$\rho(X) = \inf_a \gamma(X - aI) \qquad \text{for } X \text{ in } \mathfrak{K}_m \odot \overline{\mathfrak{K}}_m$$

where inf is taken over the set of all complex numbers a. It is readily seen that $\rho(X)$ is a quasi-norm, and that $\rho(I) = 0$. To prove that ρ possesses the cross-property it is obviously sufficient to show that $\rho(X) = \gamma(X)$ for all operators X of rank 1. We notice first that $\rho(X) \leqslant \gamma(X)$, by definition of ρ stated above. On the other hand, Lemma 5.48 gives $\rho(X) \geqslant \gamma(X)$ for all operators X of rank 1. Thus, $\rho(X) = \gamma(X)$ for all operators X of rank 1. This completes the proof.

THEOREM 5.19. Given an ε, $0 < \varepsilon \leqslant 1$, there exists a crossnorm α defined on $\mathfrak{K}_m \odot \overline{\mathfrak{K}}_m$ with $\alpha(I) = \varepsilon$.

Proof. We put,

$$\alpha(X) = (1 - \varepsilon)\,\rho(X) + \varepsilon\,\gamma(X)$$

where ρ has the meaning given in Lemma 5.49. This concludes the proof.

THEOREM 5.20. λ is not the least crossnorm on $\mathfrak{H}_n \odot \overline{\mathfrak{H}}_n$.

Proof. Since λ is a crossnorm, $\lambda(I) > 0$. Choose an ε for which $0 < \varepsilon < \lambda(I)$. By Theorem 5.19 we construct a crossnorm α for which $\alpha(I) = \varepsilon < \lambda(I)$. This concludes the proof.

THEOREM 5.21. We can construct crossnorms on $\mathfrak{H}_n \odot \overline{\mathfrak{H}}_n$ whose associates are not crossnorms.

Proof. Since λ represents the least crossnorm whose associate is also a crossnorm (Theorem 2.1), our statement is an immediate consequence of Theorem 5.20.

COROLLARY 5.4. We can construct norms on $\mathfrak{H}_n \odot \overline{\mathfrak{H}}_n$ which are not crossnorms and whose associates are crossnorms.

Proof. $\mathfrak{H}_n$ is finite-dimensional. Thus for any norm α on $\mathfrak{H}_n \odot \overline{\mathfrak{H}}_n$ we have $\alpha'' = \alpha$. Clearly any crossnorm constructed in Theorem 5.21 satisfies the required conditions.

REMARK 5.8. The arguments presented above exclude the case $n = \infty$ that is when $\mathfrak{H}_\infty$ represents a Hilbert space $\mathfrak{H}$. This is due to the fact that the identity operator I on $\mathfrak{H}_n$, belongs to (tc) only for a finite n . It is, however, not difficult to take care of the case $n = \infty$ by a modification of the above procedure in the course of which I is replaced by any projection on $\mathfrak{H}$ whose range is at least four-dimensional.

COROLLARY 5.5. None of the crossnorms constructed in Theorem 5.19 is unitarily invariant (uniform).

Proof. By Theorem 5.4, all unitarily invariant crossnorms must be $\geqslant \lambda$. This concludes the proof.

APPENDIX I

1. Reflexive crossnorms.

Let f denote a fixed element in a Banach space $\mathfrak{B}$. For $F \in \mathfrak{B}^*$ we put

$$\mathfrak{f}(F) = F(f) .$$

Then, $\mathfrak{f}$ is an additive bounded functional on $\mathfrak{B}^*$; its bound $|||\mathfrak{f}||| = \|f\|$ [1, p. 190] . The bound represents a norm on $\mathfrak{B}^{**}$. We write this $|||\mathfrak{f}||| = \|\mathfrak{f}\|$ Thus, we may assume $\mathfrak{B} \subset \mathfrak{B}^{**}$ [1, p. 180] . A space $\mathfrak{B}$ for which $\mathfrak{B} = \mathfrak{B}^{**}$ in the sense described above is termed reflexive.

An expression $\mathfrak{B}_1 \odot \mathfrak{B}_2$ may therefore be also considered as one in $\mathfrak{B}_1^{**} \odot \mathfrak{B}_2^{**}$. We write therefore,

$$\mathfrak{B}_1 \odot \mathfrak{B}_2 \subset \mathfrak{B}_1^{**} \odot \mathfrak{B}_2^{**} .$$

By Lemma 2.11, whenever α is a crossnorm $\geqq \lambda$ on $\mathfrak{B}_1 \odot \mathfrak{B}_2$, then α', α'',..... are also crossnorms on $\mathfrak{B}_1^* \odot \mathfrak{B}_2^*$, $\mathfrak{B}_1^{**} \odot \mathfrak{B}_2^{**}$,....... respectively. Since both α and α'' are defined on $\mathfrak{B}_1 \odot \mathfrak{B}_2$, we may investigate there their mutual relationship. A similar statement concerns α' and α''' on $\mathfrak{B}_1^* \odot \mathfrak{B}_2^*$.

For the sake of simplicity we shall assume throughout this Appendix that $\mathfrak{B}_1$ and $\mathfrak{B}_2$ denote two reflexive Banach spaces. This is equivalent to the statement that $\mathfrak{B}_1^*$ and $\mathfrak{B}_2^*$ are reflexive [10, p. 421] . For a slightly modified interpretation some of the statements below also hold for perfectly general

Banach spaces. We also stipulate that all crossnorms α considered are $\geq \lambda$ without having to repeat this assumption each time explicitly.

We begin with a few simple statements concerning a norm. It is easy to prove (as indicated below) that whenever at least one of the spaces $\mathcal{B}_1$, $\mathcal{B}_2$, is finite dimensional, we have always $\alpha'' = \alpha$. We have already shown (Theorem 5.5) that the last relation holds for unitarily invariant (uniform) crossnorms on $\mathcal{K} \odot \overline{\mathcal{K}}$. In the most general case we have:

LEMMA 1. $\alpha'' \leq \alpha$ on $\mathcal{B}_1 \odot \mathcal{B}_2$.

Proof. Let $\sum_{i=1}^{n} f_i \otimes g_i$ be a fixed expression in $\mathcal{B}_1 \odot \mathcal{B}_2$. We remark first that $\alpha(\sum_{i=1}^{n} f_i \otimes g_i)$ is a constant c which satisfies the inequality:

$$|(\sum_{i=1}^{n} f_i \otimes g_i)(\sum_{j=1}^{m} F_j \otimes G_j)| =$$

$$|(\sum_{j=1}^{m} F_j \otimes G_j)(\sum_{i=1}^{n} f_i \otimes g_i)| \leq c\, \alpha'(\sum_{j=1}^{m} F_j \otimes G_j)$$

for all expressions $\sum_{j=1}^{m} F_j \otimes G_j$ in $\mathcal{B}_1^* \odot \mathcal{B}_2^*$. On the other hand since by Definition 2.2, $\alpha''(\sum_{i=1}^{n} f_i \otimes g_i)$ represents the least of such constants, we have, $\alpha''(\sum_{i=1}^{n} f_i \otimes g_i) \leq \alpha(\sum_{i=1}^{n} f_i \otimes g_i)$. This concludes the proof.

REMARK 1. The last Lemma is also true for perfectly general Banach spaces, that is, we have $\alpha'' \leq \alpha$ on $\mathcal{B}_1 \odot \mathcal{B}_2 \subset \mathcal{B}_1^{**} \odot \mathcal{B}_2^{**}$.

REMARK 2. At this point one is tempted to reason as follows: For any Banach space $\mathcal{B}$ we have $\mathcal{B} \subset \mathcal{B}^{**}$. In this inclusion the norm for elements in $\mathcal{B}$ coincides with the norm of those elements when considered in $\mathcal{B}^{**}$. Would not the same argument prove that in general the norm α coincides with α''? The answer is "no" since in general $\mathcal{B}_1^{**} \otimes_{\alpha''} \mathcal{B}_2^{**}$ may not coincide with $(\mathfrak{B}_1 \otimes_\alpha \mathfrak{B}_2)^{**}$. An example readily follows from Theorem 5.15 where $\mathfrak{H} \otimes_\lambda \overline{\mathfrak{H}} = \mathfrak{H} \otimes_{\lambda'} \overline{\mathfrak{H}}$ represents the space of completely continuous operators on $\mathfrak{H}$, while $(\mathfrak{H} \otimes_\lambda \overline{\mathfrak{H}})^{**}$ is the space of all operators on $\mathfrak{H}$. In fact, in the general case for a given cross-space, its conjugate space contains the associate space as a proper subspace. The precise conditions for their equality are stated in Theorem 3.6. Lemma 4 below proves that whenever the conjugate and associate space coincide, we actually have $\alpha'' = \alpha$.

LEMMA 2. $\alpha''' = \alpha'$ on $\mathfrak{B}_1^* \odot \mathfrak{B}_2^*$.

Proof. By Lemma 1, $\alpha'' \leq \alpha$. Thus, Lemma 2.2 implies $\alpha''' \geq \alpha'$. Now applying Lemma 1 to α' instead of α we get $\alpha''' \leq \alpha'$. Thus, $\alpha''' = \alpha'$. This concludes the proof.

DEFINITION 1. A crossnorm α on $\mathfrak{B}_1 \odot \mathfrak{B}_2$ will be termed

(i) minimal, if for every norm β on $\mathfrak{B}_1 \odot \mathfrak{B}_2$ for which $\alpha' = \beta'$ we have $\alpha \leq \beta$.

(ii) reflexive, if $\alpha'' = \alpha$.

(iii) having an associate property, if for some crossnorm β on $\mathfrak{B}_1^* \odot \mathfrak{B}_2^*$ we have $\alpha = \beta'$.

THEOREM 1. For a crossnorm α on $\mathfrak{B}_1 \odot \mathfrak{B}_2$ the following statements are equivalent:

(i) α is minimal

(ii) α is reflexive

(iii) α has an associate property.

Proof. We shall prove (i)$\to$(ii)$\to$(iii)$\to$(i).

We assume (i). By Lemma 2, $\alpha''' = \alpha'$. Thus, α'' and α have the same associate α'. By assumption α is minimal. Thus, $\alpha \leq \alpha''$. On the other hand Lemma 1 gives $\alpha'' \leq \alpha$. This proves (i)$\to$(ii).

We assume $\alpha = \alpha''$. Then clearly α is the associate of α'. Thus, (ii)$\to$(iii).

Finally, we assume (iii). This means that for some crossnorm β on $\mathfrak{B}_1^* \odot \mathfrak{B}_2^*$ we have $\alpha = \beta'$. Let α_0 denote a crossnorm on $\mathfrak{B}_1 \odot \mathfrak{B}_2$ for which $\alpha_0' = \alpha'$. By Lemmas 1 and 2, $\alpha_0 \geqslant \alpha_0'' = \alpha'' = \beta''' = \beta' = \alpha$. Thus, $\alpha_0 \geqslant \alpha$, that is, (iii)$\to$(i). This concludes the proof.

LEMMA 3. λ is a reflexive crossnorm.

Proof. By Lemma 2.11, $\lambda'' \geqslant \lambda$. On the other hand Lemma 1 gives $\lambda'' \leq \lambda$. Thus, $\lambda'' = \lambda$.

REMARK 3. The last Lemma is also valid for perfectly general Banach spaces. This follows from the following argument: By Lemma 2.12, λ on $\mathfrak{B}_1^{**} \odot \mathfrak{B}_2^{**}$ is an extension of λ on $\mathfrak{B}_1 \odot \mathfrak{B}_2$. By Lemma 2.11, $\lambda'' \geqslant \lambda$ on $\mathfrak{B}_1^{**} \odot \mathfrak{B}_2^{**}$ hence also on $\mathfrak{B}_1 \odot \mathfrak{B}_2$. An application of Lemma 1, which holds for perfectly general Banach spaces, proves our contention.

LEMMA 4. Whenever for a cross-space $\mathfrak{B}_1 \otimes_\alpha \mathfrak{B}_2$ its conjugate space coincides with its associate space, we have, $\alpha'' = \alpha$.

Proof. We recall first that for a fixed f in a Banach space $\mathfrak{B}$, $\| f \|$ is the least number c satisfying the inequality

$$|F(f)| \leq c\,\|F\| \quad \text{for all} \quad F \in \mathcal{B}^*, \quad (\|F\| = |||F|||)\,.$$

Since a norm is clearly a continuous function, the value of the last c will be unaffected if we restrict the F's only to a dense set in $\mathcal{B}^*$.

By assumption $\mathcal{B}_1^* \odot_{\alpha'} \mathcal{B}_2^*$ is dense in $(\mathcal{B}_1 \otimes_\alpha \mathcal{B}_2)^*$. Hence if $\sum_{i=1}^{n} f_i \otimes g_i$ and $\mathcal{B}_1 \otimes_\alpha \mathcal{B}_2$ take the place of f and $\mathcal{B}$ in our previous remark, we see that $\alpha(\sum_{i=1}^{n} f_i \otimes g_i)$ represents the least constant c satisfying the inequality

$$\left|\left(\sum_{j=1}^{m} F_j \otimes G_j\right)\left(\sum_{i=1}^{n} f_i \otimes g_i\right)\right| \leq c\,\alpha'\left(\sum_{j=1}^{m} F_j \otimes G_j\right)$$

for all $\sum_{j=1}^{m} F_j \otimes G_j$ in $\mathcal{B}_1^* \odot_{\alpha'} \mathcal{B}_2^*$. By Definition 2.2 however, that constant is $\alpha''(\sum_{i=1}^{n} f_i \otimes g_i)$. This concludes the proof.

2. Reflexive cross-spaces.

Here we shall say a few words about reflexive cross-spaces. In particular, we point out the relationship between such cross-spaces and those for which the conjugate and the associate space are identical.

Suppose the cross-space $\mathcal{B}_1 \otimes_\alpha \mathcal{B}_2$ is reflexive. Clearly, $\mathcal{B}_1$ and $\mathcal{B}_2$ may be considered closed linear subspaces in it. Since a closed linear subspace of a reflexive Banach space is also reflexive [10, p. 423], both, $\mathcal{B}_1$ and $\mathcal{B}_2$ must be reflexive. The converse statement is not true. Both $\mathcal{B}_1$ and $\mathcal{B}_2$ may be reflexive and in addition $\alpha'' = \alpha$, while $\mathcal{B}_1 \otimes_\alpha \mathcal{B}_2$ may not be reflexive. Although such examples can be easily constructed in the light of Theorem 2 below, we prefer to outline the details of the following one directly, [4, p. 433].

EXAMPLE. Let $p > 1$. We denote by l_p the Banach space of all sequences of real numbers $\{x_i\}$ for which $\sum_i |x_i|^p < +\infty$ and where

$(\sum_i |x_i|^p)^{1/p}$ represents the norm of $\{x_i\}$. It is well-known that l_p^* may be interpreted as l_q where $\frac{1}{p} + \frac{1}{q} = 1$. Thus, $l_p^{**} = l_p$ [1, p. 68].

Let $p > 1$. Then, $l_p \otimes_\lambda l_q$ is non-reflexive.

Proof. Let $\Phi_1, \Phi_2, \Phi_3, \ldots$ and $\varphi_1, \varphi_2, \varphi_3, \ldots$ denote the sequence of elements $(1, 0, 0, \ldots)$, $(0, 1, 0, \ldots)$, $(0, 0, 1, \ldots)$ in l_p and l_q respectively. We have:

$$\Phi_i(\varphi_j) = 0 \quad \text{if } i \neq j \quad \text{and} \quad \Phi_i(\varphi_i) = 1 \; ; \; i, j = 1, 2, \ldots$$

Let $n > m$. For any constant real numbers $a_m, a_{m+1}, \ldots, a_n$, we have,

$$\lambda\left(\sum_{i=m}^{n} a_i \Phi_i \otimes \varphi_i\right) = \max_{m \leq i \leq n} |a_i| .$$

By Definition 2.4, $\lambda(\sum_{i=m}^{n} a_i \Phi_i \otimes \varphi_i) = \sup \left| \sum_{i=m}^{n} a_i \Phi_i(\varphi) \Phi(\varphi_i) \right|$ where sup is taken over all $\Phi \in l_p$, $\varphi \in l_q$, such that $\|\Phi\| = \|\varphi\| = 1$. Putting $\Phi = \Phi_i$ and $\varphi = \varphi_i$ in succession for $m \leq i \leq n$, we get

$$\lambda\left(\sum_{i=m}^{n} a_i \Phi_i \otimes \varphi_i\right) \geq \max_{m \leq i \leq n} |a_i| .$$

On the other hand, for $\Phi \in l_p$ with $\Phi = x_1 \Phi_1 + x_2 \Phi_2 + \ldots$ we have $\|\Phi\| = 1$ if and only if, $\sum_i |x_i|^p = 1$. Similarly, if $\varphi \in l_q$ and $\varphi = y_1 \varphi_1 + y_2 \varphi_2 + \ldots$, then $\|\varphi\| = 1$ if and only if, $\sum_i |y_i|^q = 1$. Thus, $\lambda(\sum_{i=m}^{n} a_i \Phi_i \otimes \varphi_i) = \sup \left| \sum_{i=m}^{n} a_i x_i y_i \right|$ with the sup extended over the set of all sequences of real numbers $\{x_i\}$, $\{y_i\}$, for which $\sum_i |x_i|^p = \sum_i |y_i|^q = 1$. Hőlder's inequality gives:

$$\lambda\left(\sum_{i=m}^{n} a_i \Phi_i \otimes \varphi_i\right) \leq \sup \left(\max_{m \leq i \leq n} |a_i|\right) \left(\sum_{i=m}^{n} |x_i|^p\right)^{\frac{1}{p}} \left(\sum_{i=m}^{n} |y_i|^q\right)^{\frac{1}{q}} .$$

The right side is clearly, $= \max_{m \leq i \leq n} |a_i|$. This concludes the proof of our equality.

Thus, the sequence of expressions $a_1\Phi_1 \otimes \varphi_1$, $\Sigma_{i=1}^{2} a_i\Phi_i \otimes \varphi_i$, $\Sigma_{i=1}^{3} a_i\Phi_i \otimes \varphi_i$, ... is fundamental if and only if, $\{a_i\} \to 0$. Whenever this is the case, its norm is clearly

$$\sup_i |a_i| .$$

Thus, the well-known non-reflexive space c_0 [1, p. 181] and [1, pp. 66-67] of all converging towards 0 sequences of real numbers may be considered a subspace of $l_p \otimes_\lambda l_q$. The last therefore must not be reflexive by [10, p. 423]. This concludes the proof.

LEMMA 5. We assume that the Banach spaces $\mathfrak{B}_1$, $\mathfrak{B}_2$, are reflexive and $\alpha'' = \alpha$. Then, the associate space forms a "fundamental" subspace of the conjugate space. We mean hereby, that whenever $\tilde{F}(\tilde{f}_0) = 0$ for all $\tilde{F}$ in $\mathfrak{B}_1^* \otimes_{\alpha'} \mathfrak{B}_2^*$ and some $\tilde{f}_0$ in $\mathfrak{B}_1 \otimes_\alpha \mathfrak{B}_2$, then $\tilde{f}_0 = 0$.

Proof. The meaning of the symbol $\tilde{F}(\tilde{f}_0)$ is obviously the following: Let $\Sigma_{j=1}^{p_m} F_j^{(m)} \otimes G_j^{(m)}$ be a fundamental sequence of expressions in $\mathfrak{B}_1^* \otimes_{\alpha'} \mathfrak{B}_2^*$ for $\tilde{F}$ and similarly $\Sigma_{i=1}^{q_m} f_i^{(m)} \otimes g_i^{(m)}$ a fundamental sequence in $\mathfrak{B}_1 \otimes_\alpha \mathfrak{B}_2$ for $\tilde{f}_0$. Then, the sequence of corresponding inner products $(\Sigma_{j=1}^{p_m} F_j^{(m)} \otimes G_j^{(m)})(\Sigma_{i=1}^{q_m} f_i^{(m)} \otimes g_i^{(m)})$ is always convergent, to the same limit, independent on the fundamental sequences of expressions representing $\tilde{F}$ and $\tilde{f}_0$. We denote this limit by $\tilde{F}(\tilde{f}_0)$

By Definition 2.2 and continuity of a norm, $\alpha''(\tilde{f}_0)$ is the least number c satisfying the inequality $|\tilde{F}(\tilde{f}_0)| \leq c\, \alpha'(\tilde{F})$ for all $\tilde{F}$ in $\mathfrak{B}_1^* \otimes_{\alpha'} \mathfrak{B}_2^*$. By assumption, $\tilde{F}(\tilde{f}_0)$ is always 0. Thus, $\alpha''(\tilde{f}_0) = 0$. Hence, $\alpha(\tilde{f}_0) = 0$ and $\tilde{f}_0 = 0$. This concludes the proof.

LEMMA 6. Whenever for a reflexive crossnorm α the space $\mathfrak{B}_1 \otimes_\alpha \mathfrak{B}_2$ is reflexive, its conjugate and associate space coincide.

Proof. Suppose to the contrary that an element $\tilde{F}_0$ of the conjugate space is not in the associate space. By construction, the associate space is "absolutely closed", that is, closed in any metric set containing it, hence also in the conjugate space. By [1, p. 57] there exists an additive bounded functional $\mathfrak{F}$ on the conjugate space such that $\mathfrak{F}(\tilde{F}_0) = 1$ and $\mathfrak{F}(\tilde{F}) = 0$ for all $\tilde{F}$ in the associate space. By assumption, our cross-space is reflexive. Thus, it contains an element $\tilde{f}$ corresponding to $\mathfrak{F}$, with $\tilde{F}_0(\tilde{f}) = 1$ and $\tilde{F}(\tilde{f}) = 0$ for all $\tilde{F}$ in the associate space. The last implies $\tilde{f} = 0$ by Lemma 5. This obviously contradicts $\tilde{F}_0(\tilde{f}) = 1$. This concludes the proof.

THEOREM 2. For a reflexive crossnorm α the following statements are equivalent:

(i) $\mathfrak{B}_1 \otimes_\alpha \mathfrak{B}_2$ is reflexive.

(ii) $\mathfrak{B}_1^* \otimes_{\alpha'} \mathfrak{B}_2^*$ is reflexive.

(iii) $(\mathfrak{B}_1 \otimes_\alpha \mathfrak{B}_2)^* = \mathfrak{B}_1^* \otimes_{\alpha'} \mathfrak{B}_2^*$.

(iii') $(\mathfrak{B}_1^* \otimes_{\alpha'} \mathfrak{B}_2^*)^* = \mathfrak{B}_1 \otimes_\alpha \mathfrak{B}_2$.

Proof. In our proof we shall use the following propositions stated in [10, pp. 421 and 423] :

(a) A Banach space $\mathfrak{B}$ is reflexive if and only if, $\mathfrak{B}^*$ is reflexive.

(b) A closed linear manifold in a reflexive Banach space is also reflexive.

We shall prove that (i) → (ii) → (iii) and (iii'), while (iii) in conjunction with (iii') implies (i).

We assume (i). By (a), $(\mathcal{B}_1 \otimes_\alpha \mathcal{B}_2)^*$ is also reflexive. Hence, its closed linear subspace $\mathcal{B}_1^* \otimes_{\alpha'} \mathcal{B}_2^*$ must be also reflexive by (b). Thus, (i) → (ii).

We assume (ii). $\mathcal{B}_1^*$, $\mathcal{B}_2^*$, may obviously be considered as closed linear subspaces in $\mathcal{B}_1^* \otimes_{\alpha'} \mathcal{B}_2^*$; they must be reflexive by (b). By Lemma 2, α' is always reflexive. Thus, replacing in Lemma 6, α by α', $\mathcal{B}_1$ by $\mathcal{B}_1^*$ and $\mathcal{B}_2$ by $\mathcal{B}_2^*$ we obtain $(\mathcal{B}_1^* \otimes_{\alpha'} \mathcal{B}_2^*)^* = \mathcal{B}_1^{**} \otimes_{\alpha''} \mathcal{B}_2^{**} = \mathcal{B}_1 \otimes_\alpha \mathcal{B}_2$. Thus, (iii') holds. To prove (iii), we apply $*$ to (take the conjugate of) both sides of (iii'), and apply (ii) to the left side of the resulting equality. Thus, (ii) → (iii) and (iii').

Finally, we assume that (iii) and (iii') hold. We apply $*$ to both sides of (iii). The resulting equality together with (iii') furnishes (i). This concludes the proof.

COROLLARY 2. For a Hilbert space $\mathcal{H}$, both $\mathcal{H} \otimes_\lambda \overline{\mathcal{H}}$ and $\mathcal{H} \otimes_\gamma \overline{\mathcal{H}}$ are non-reflexive.

Proof. By Lemma 3, λ is reflexive. Furthermore, by Corollary 5.2, $\mathcal{H} \otimes_\lambda \overline{\mathcal{H}} = \mathcal{H} \otimes_{\gamma'} \overline{\mathcal{H}}$ is a proper subspace of $(\mathcal{H} \otimes_\gamma \overline{\mathcal{H}})^*$. Thus, (iii') of Theorem 2 does not hold. Consequently, neither (i) or (ii) of Theorem 2 holds. This concludes the proof.

COROLLARY 3. For a reflexive crossnorm α, the space $\mathcal{B}_1 \otimes_\alpha \mathcal{B}_2$ is reflexive if and only if, every operator from $\mathcal{B}_1$ into $\mathcal{B}_2^*$ of finite α-norm can be approximated in that norm by operators of finite rank and every operator from $\mathfrak{B}_2^*$ into $\mathfrak{B}_1$ of finite α'-norm can be approximated in that norm by operators of finite rank.

Proof. The proof is a consequence of Theorems 3.6 and 2.

It does not appear to be a simple matter to state some "reasonable" sufficient conditions for a crossnorm for which the resulting cross-space is such that its conjugate space coincides with its associate space. The theorem which follows may be of interest although the type of condition stated herein is far from being reasonable.

THEOREM 3. For a uniformly convex [2b] crossnorm α on $\mathfrak{B}_1 \odot \mathfrak{B}_2$, the conjugate space coincides with the associate space if and only if, α is reflexive.

Proof. Let α be uniformly convex and $\alpha'' = \alpha$. By continuity, α is also uniformly convex on $\mathfrak{B}_1 \otimes_\alpha \mathfrak{B}_2$. The last cross-space must be reflexive since it is known that a Banach space with a uniformly convex norm is reflexive. Thus, by Theorem 2, its conjugate space must coincide with its associate space. That the converse holds for any crossnorm is the content of Lemma 4. This concludes the proof.

3. "Limited" crossnorms.

We conclude this Appendix with a few remarks about a class of reflexive crossnorms which we shall term "limited". The notion of such a crossnorm on $\mathfrak{H} \odot \bar{\mathfrak{H}}$ has already been mentioned before (Definition 5.10) since in that place we have found it convenient to discuss the topological equivalence of the spaces $\mathfrak{H}\otimes_\lambda\overline{\mathfrak{H}}$ and $\mathfrak{H}\otimes_\alpha\overline{\mathfrak{H}}$ for any limited crossnorm α.

DEFINITION 2. Let $\mathfrak{B}_1$ and $\mathfrak{B}_2$ denote any two Banach spaces and α a crossnorm on $\mathfrak{B}_1^*\odot\mathfrak{B}_2^*$. We define a sequence of functions $\{\alpha_p\}$ on $\mathfrak{B}_1\odot\mathfrak{B}_2$ as follows: $\alpha_p(\Sigma_{i=1}^m f_i\otimes g_i)$ is the least constant c satisfying the inequality

$$|(\Sigma_{j=1}^p F_j\otimes G_j)(\Sigma_{i=1}^m f_i\otimes g_i)| \leq c\,\alpha(\Sigma_{j=1}^p F_j\otimes G_j)$$

for all expressions $\Sigma_{j=1}^p F_j\otimes G_j$ of rank $\leq$ p in $\mathfrak{B}_1^*\odot\mathfrak{B}_2^*$.

Thus, every crossnorm α on $\mathfrak{B}_1^*\odot\mathfrak{B}_2^*$ generates a corresponding sequence of functions $\{\alpha_p\}$ on $\mathfrak{B}_1\odot\mathfrak{B}_2$.

THEOREM 4.

(i) All α_p are crossnorms.

(ii) $\alpha_1 \leq \alpha_2 \leq \ldots$.

(iii) $\alpha_1 = \lambda$.

(iv) $\lim_{p\to\infty} \alpha_p = \alpha'$.

(v) All α_p are reflexive.

Proof. (ii) and (iii) are a consequence of Definition 2. From Definition 2. it is also clear that the associate α' is not smaller than each α_p. Thus, $\lambda\leq\alpha_p\leq\alpha'$ or α_p possesses the cross-property. The verification that α_p is a norm presents no difficulty. This proves (i). Clearly, (iv) is an immediate consequence of the definitions of α' and α_p for a given α. It remains to prove (v). From the definition of α_p for an expression $\Sigma_{i=1}^m f_i\otimes g_i$ we have,

$$\left|(\Sigma_{j=1}^{p} F_j \otimes G_j)(\Sigma_{i=1}^{m} f_i \otimes g_i)\right| \leqslant \alpha(\Sigma_{j=1}^{p} F_j \otimes G_j)\ \alpha_p(\Sigma_{i=1}^{m} f_i \otimes g_i)$$

Thus,

$$(1) \qquad (\alpha_p)'(\Sigma_{j=1}^{p} F_j \otimes G_j) \leqslant \alpha(\Sigma_{j=1}^{p} F_j \otimes G_j)$$

for all expressions of rank $\leqslant p$. Now, for a given $\Sigma_{i=1}^{m} f_i \otimes g_i$ Definition 2.2 gives,

$$\left|(\Sigma_{j=1}^{p} F_j \otimes G_j)(\Sigma_{i=1}^{m} f_i \otimes g_i)\right| \leqslant (\alpha_p)'(\Sigma_{j=1}^{p} F_j \otimes G_j)\ (\alpha_p)''(\Sigma_{i=1}^{m} f_i \otimes g_i)$$

By (1), the right side is

$$\leqslant \alpha(\Sigma_{j=1}^{p} F_j \otimes G_j)\ (\alpha_p)''(\Sigma_{i=1}^{m} f_i \otimes g_i) \quad ,$$

for all expressions $\Sigma_{j=1}^{p} F_j \otimes G_j$ of rank $\leqslant p$. This, by Definition 2 implies $\alpha_p(\Sigma_{i=1}^{m} f_i \otimes g_i) \leqslant (\alpha_p)''(\Sigma_{i=1}^{m} f_i \otimes g_i)$. Thus, $\alpha_p \leqslant (\alpha_p)''$. An application of Lemma 1 concludes the proof.

THEOREM 5. Let α be a reflexive crossnorm. Then, α_p is the least crossnorm whose associate $(\alpha_p)'$ coincides with α for all expressions of rank $\leqslant p$.

Proof. Formula (1) of Theorem 4 (v), proves that $(\alpha_p)' \leqslant \alpha$ for all expressions of rank $\leqslant p$. Again by Theorem 4 (ii), $\alpha_p \leqslant \alpha'$ and hence $(\alpha_p)' \geqslant \alpha'' = \alpha$. Thus, $(\alpha_p)' = \alpha$ for all expressions of rank $\leqslant p$.

Now suppose that for a crossnorm β and an expression $\Sigma_{i=1}^{m} f_i \otimes g_i$ we have, $\beta(\Sigma_{i=1}^{m} f_i \otimes g_i) < \alpha_p(\Sigma_{i=1}^{m} f_i \otimes g_i)$. By Definition 2, there must exist an expression $\Sigma_{j=1}^{p} F_j \otimes G_j$ of rank $\leqslant p$ for which

$$\beta(\Sigma_{i=1}^{m} f_i \otimes g_i)\ \alpha(\Sigma_{j=1}^{p} F_j \otimes G_j) < \left|(\Sigma_{j=1}^{p} F_j \otimes G_j)(\Sigma_{i=1}^{m} f_i \otimes g_i)\right| \quad .$$

By Definition 2.2, the right side is $\leqslant \beta'(\Sigma_{j=1}^{p} F_j \otimes G_j)\ \beta(\Sigma_{i=1}^{m} f_i \otimes g_i)$, and therefore $\alpha(\Sigma_{j=1}^{p} F_j \otimes G_j) < \beta'(\Sigma_{j=1}^{p} F_j \otimes G_j)$.

Thus, $\beta' = \alpha$ for all expressions of rank $\leq$ p implies $\beta \geq \alpha_p$. This concludes the proof.

REMARK 4. It is not difficult to see that the last theorem is a generalization of Theorem 2.1.

We conclude this section by pointing out some interesting properties of the "limited" crossnorms generated by the unique self-associate crossnorm σ on $\mathfrak{H} \odot \overline{\mathfrak{H}}$ (Definition 5.9 and Lemma 5.39).

LEMMA 7. Let α be a unitarily invariant crossnorm on $\mathfrak{H} \odot \overline{\mathfrak{H}}$. Let X be a fixed operator on $\mathfrak{H}$ of finite rank, whose range spans a linear manifold $\mathfrak{M}$. Then,

$$\alpha_p(X) = \sup_{\mathfrak{R}(Y) \subset \mathfrak{M}} \frac{|(X, Y)|}{\alpha(Y)}$$

where the last sup is restricted to the set of all those operators Y on $\mathfrak{H}$ of rank $\leq$ p whose range $\mathfrak{R}(Y)$ is included in $\mathfrak{M}$.

Proof. Let P be the projection of $\mathfrak{H}$ on $\mathfrak{M}$. By Lemma 5.9 (vi),

$$(X, Y) = (PX, Y) = (X, P^*Y) = (X, PY) .$$

By Lemma 5.35, the unitary invariance of α implies its uniformity. Thus,

$$\alpha(PY) \leq |||P||| \; \alpha(Y) = \alpha(Y) .$$

The last two relationships and the definition of $\alpha_p(X)$ furnish the proof.

We consider the sequence of limited crossnorms $\{\sigma_p\}$ corresponding to the unique self-associate crossnorm σ.

LEMMA 8. For every natural p, $\sigma_p \neq \sigma_{p+1}$.

Proof. Let $\varphi_1, \ldots, \varphi_p, \varphi_{p+1}$ be an orthonormal set. Using Lemma 7 we readily verify

$$\sigma_p(\Sigma_{i=1}^{p+1} \varphi_i \otimes \overline{\varphi}_i) = \sqrt{p} \quad \text{and} \quad \sigma_{p+1}(\Sigma_{i=1}^{p+1} \varphi_i \otimes \overline{\varphi}_i) = \sqrt{p+1}.$$

This concludes the proof.

COROLLARY 4. $\sigma_p \neq \sigma$ for $p = 1, 2, \ldots$.

Proof. We have $\sigma_p \leq \sigma_{p+1} \leq \sigma' = \sigma$. But $\sigma_p \neq \sigma_{p+1}$. Thus, $\sigma_p \neq \sigma$.

THEOREM 6. For every natural p, we have $\sigma_p = (\sigma_p)' = \sigma$ for all expressions of rank $\leq p$.

Proof. Let α be a crossnorm. Then, $(\sigma(X))^2 = (X, X) \leq \alpha(X)\, \alpha'(X)$. Thus, for any crossnorm α we have $\sigma^2 \leq \alpha\alpha'$ and in particular $\sigma^2 \leq \sigma_p(\sigma_p)'$. Now, $\sigma_p \leq \sigma' = \sigma$ implies $(\sigma_p)' \geq \sigma' = \sigma$. By Theorem 5, $(\sigma_p)' = \sigma$ for all expressions of rank $\leq p$. Hence also $\sigma_p = \sigma$ for all expressions of rank $\leq p$. This concludes the proof.

REMARK 5. From Theorem 6 and 4(v) it is clear that we have constructed three different reflexive crossnorms which are equal to each other for all expressions of rank $\leq p$, namely, σ_p, $(\sigma_p)'$, σ . The crossnorms σ_p and $(\sigma_p)'$ are associate with each other. Since, they are also equal for all expressions of rank $\leq p$ we may term them "semi-self-associate".

REMARK 6. From Remark 5 also follows that a crossnorm on $\mathfrak{H} \otimes \overline{\mathfrak{H}}$ is not determined by the values it assumes for all expressions of rank $\leq p$ (where p is any natural number smaller than the dimension of $\mathfrak{H}$) .

1. A self-associate crossnorm.

THEOREM. It is possible to define a construction which for any two Banach spaces $\mathfrak{B}_1$, $\mathfrak{B}_2$, (without any special restrictions!) determines a crossnorm on $\mathfrak{B}_1 \odot \mathfrak{B}_2$. Moreover, this construction when applied to two unitary spaces, furnishes the usual unique self-associate crossnorm σ (Definition 5.1 and Lemma 5.39).

We precede our proof with the following two Lemmas:

LEMMA 1. For positive numbers a, b, we have,

$$\frac{1}{a+b} \leq \frac{1}{4a} + \frac{1}{4b} .$$

Proof. The proof is immediate.

LEMMA 2. For any two norms α, β, we have

$$\left(\frac{\alpha+\beta}{2}\right)' \leq \frac{\alpha'+\beta'}{2} .$$

Proof. Let $\tilde{F} \in \mathfrak{B}_1^* \odot \mathfrak{B}_2^*$ be fixed. By Lemma 1, for any non-zero $\tilde{f}$ in $\mathfrak{B}_1 \odot \mathfrak{B}_2$ we have,

$$\frac{2\,|\tilde{F}(\tilde{f})|}{\alpha(\tilde{f}) + \beta(\tilde{f})} \leq \frac{1}{2}\left(\frac{|\tilde{F}(\tilde{f})|}{\alpha(\tilde{f})} + \frac{|\tilde{F}(\tilde{f})|}{\beta(f)} \right) .$$

Thus, Definition 2.2 furnishes the proof.

We are ready for the proof of our Theorem. Instead of following the usual pattern, we use a more suggestive approach with the introduction of the notion of a "general crossnorm". We do not attempt here to give a precise formulation of this notion. For our purpose it is sufficient to remark that the greatest crossnorm γ is such, since it is uniquely defined on $\mathfrak{B}_1 \odot \mathfrak{B}_2$ for any two Banach spaces $\mathcal{B}_1$, $\mathcal{B}_2$. Similarly, λ as well as $\frac{\gamma+\lambda}{2}$ are general crossnorms. When α is a general crossnorm, α' should also stand for a general crossnorm with the following significance on $\mathcal{B}_1 \odot \mathcal{B}_2$: We consider α on $\mathcal{B}_1^* \odot \mathcal{B}_2^*$. This determines α' on $\mathcal{B}_1^{**} \odot \mathcal{B}_2^{**}$. The latter we confine to $\mathcal{B}_1 \odot \mathcal{B}_2 \subset \mathcal{B}_1^{**} \odot \mathcal{B}_2^{**}$. Thus, γ, γ', λ, λ', $\frac{\gamma+\lambda}{2}$, $\left(\frac{\gamma+\lambda}{2}\right)'$, ... are general crossnorms.

We proceed with our construction:

Put $\alpha_1 = \frac{\gamma+\gamma'}{2}$ and $\alpha_n = \frac{\alpha_{n-1} + (\alpha_{n-1})'}{2}$ for $n > 1$.

By Lemma 2,
$$(\alpha_n)' = \left(\frac{\alpha_{n-1} + (\alpha_{n-1})'}{2}\right)' \leqslant \frac{(\alpha_{n-1})' + (\alpha_{n-1})''}{2}$$
$$\leqslant \frac{(\alpha_{n-1})' + \alpha_{n-1}}{2} = \alpha_n .$$

Thus,
$$\gamma' \leqslant (\alpha_1)' \leqslant (\alpha_2)' \leqslant \;.... \quad ... \leqslant \alpha_2 \leqslant \alpha_1 \leqslant \gamma .$$

Put, $\lim_{n\to\infty} \alpha_n = \alpha$ and $\lim_{n\to\infty} (\alpha_n)' = \beta$.

Since,
$$\alpha_1 - (\alpha_1)' = \tfrac{1}{2}(\gamma+\gamma') - (\alpha_1)' \leqslant \tfrac{1}{2}(\gamma+\gamma') - \gamma' = \tfrac{1}{2}(\gamma-\gamma').$$
$$\alpha_2 - (\alpha_2)' = \tfrac{1}{2}\left(\alpha_1 + (\alpha_1)'\right) - (\alpha_2)' \leqslant \tfrac{1}{2}\left(\alpha_1 + (\alpha_1)'\right) - (\alpha_1)' =$$
$$\tfrac{1}{2}\left(\alpha_1 - (\alpha_1)'\right) \leqslant \tfrac{1}{2^2}(\gamma - \gamma'),$$

and in general,
$$\alpha_n - (\alpha_n)' \leqslant \frac{1}{2^n}(\gamma - \gamma'),$$

we have, $\alpha - \beta \equiv 0$ or $\alpha = \beta$.

Since, $\alpha \leqslant \alpha_n$ and consequently $\alpha' \geqslant (\alpha_n)'$ for all n, we have $\alpha' \geqslant \beta = \alpha$.

Since $(\alpha_n)' \leqslant \beta = \alpha$ hence, $\alpha_n \geqslant (\alpha_n)'' \geqslant \alpha'$ for all n, we have, $\alpha \geqslant \alpha'$.

Thus, $\alpha = \alpha'$.

The obtained general crossnorm α is defined for any two Banach spaces. It is clear that for the case of a unitary space $\mathfrak{K}$ the resulting α coincides with α' on $\mathfrak{K} \odot \overline{\mathfrak{K}}$, hence, is self-associate in the sense of Definition 5.9. By Lemma 5.39, α coincides with σ.

REFERENCES

1. S. Banach,
   Théorie des opérations linéaires, Warsaw, 1932.

2. T. Bonnesen and W. Fenchel,
   Theorie der konvexen Körper
   (Ergebnisse der Math. vol. III) J. Springer, Berlin, 1934.

2a. J. A. Clarkson,
   Uniformly Convex Spaces
   Trans. Amer. Math. Soc., vol. 40 (1936) pp. 396-414.

2b. J. W. Calkin,
   Two Sided Ideals and Congruences
   Annals of Math., vol. 42 (1941) pp. 839-873.

3. J. L. Dorroh,
   Concerning the Direct Product of Algebras
   Annals of Math., vol. 36 (1935) pp. 882-885.

4. N. Dunford and R. Schatten,
   On the Associate and Conjugate Space for the Direct Product
   Trans. Amer. Math. Soc., vol. 59 (1946) pp. 430-436.

5. F. Hausdorff,
   Mengenlehre, Walter de Gruyter, Berlin, 1935.

6. S. Kakutani,
   Some Characterizations of Euclidean Space
   Japanese Journal of Math. vol. 16, (1939) pp. 93-97.

7. F. J. Murray,
   On Complementary Manifolds and Projections in Spaces L_p and l_p
   Trans. Amer. Math. Soc., vol. 41 (1937) pp. 138-152.

8. F. J. Murray,
   Bilinear Transformations in Hilbert Space
   Trans. Amer. Math. Soc., vol. 45 (1939) pp. 474-507.

9. F. J. Murray and J. von Neumann,
   On Rings of Operators
   Annals of Math. (2) vol. 37 (1936) pp. 116-229.

10. B. J. Pettis,
   A Note on Regular Banach Spaces
   Bull. Amer. Math. Soc., vol. 44 (1938) pp. 420-428.

11. R. S. Phillips,
On Linear Transformations
Trans. Amer. Math. Soc., vol. 48 (1940) pp. 516-541.

12. R. Schatten,
On the Direct Product of Banach Spaces
Trans. Amer. Math. Soc., vol. 53 (1943) pp. 195-217.

13. R. Schatten,
On Reflexive Norms for the Direct Product
Trans. Amer. Math. Soc., vol. 54 (1943) pp. 498-506.

14. R. Schatten,
The Cross-Space of Linear Transformations
Annals of Math., vol. 47 (1946) pp. 73-84.

15. R. Schatten and J. von Neumann,
The Cross-Space of Linear Transformations II
Annals of Math., vol. 47 (1946) pp. 608-630.

16. R. Schatten and J. von Neumann,
The Cross-Space of Linear Transformations III
Annals of Math., vol. 49 (1948) pp. 557-582.

17. R. Schatten,
On Projections with Bound 1
Annals of Math., vol. 48 (1947) pp. 321-325.

17a. R. Schatten,
"Closing-up" of Sequence Spaces
Amer. Math. Monthly, to be published in 1950.

18. M. H. Stone,
Linear Transformations in Hilbert Space
Amer. Math. Soc., Colloquium Publications, vol. XV (1932).

19. J. von Neumann,
On Infinite Direct Products
Compositio Math., vol. 6 (1938) pp. 1-77.

20. J. von Neumann,
Mathematische Begründung der Quantenmechanic
Göttinger Nachrichten (1927) pp. 1-54.

21. J. von Neumann,
Mathematische Grundlagen der Quantenmechanic
Die Grundlehren der Mathematischen Wissenschaften, Berlin, 1932,
or Dover Publication, New York, 1943.

22. J. von Neumann,
Uber adjungierte Funktionaloperatoren,
Annals of Math., vol. 33 (1932) pp. 294-310.

23. J. von Neumann,
Some Matrix-Inequalities and Metrization of Matrix-Space
Tomsk University Review, vol. I (1937) pp. 286-300.

24. J. von Neumann,
Allgemeine Eigenwerttheorie Hermitescher Funktionaloperatoren
Math. Annalen, vol. 102 (1929) pp. 49-131.

25. H. Weyl,
The Theory of Groups and Quantum-mechanics
Translated from German by H. P. Robertson, New York, 1931.

26. H. Whitney,
Tensor Products of Abelian Groups
Duke Math. Journal, vol. 4 (1938) pp. 495-528.

Ingram Content Group UK Ltd.
Milton Keynes UK
UKHW012247050523
421313UK00001B/81